CARROLL COLLEGE LIBRARY
HELENA, MONTANA 59625

DISCARD

Scarce Water and
Institutional Change

Scarce Water and Institutional Change

Kenneth D. Frederick, editor

*with the assistance of
Diana C. Gibbons*

PUBLISHED BY RESOURCES FOR THE FUTURE, INC.
WASHINGTON, D.C.

Copyright ©1986 by Resources for the Future, Inc.
All rights reserved
Manufactured in the United States of America

Published by Resources for the Future, Inc.
1616 P Street, N.W., Washington, D.C. 20036
Resources for the Future books are distributed worldwide by
The Johns Hopkins University Press

Library of Congress Cataloging-in-Publication Data
Main entry under title:

Scarce water and institutional change.

  Includes index.
  1. Water-supply—United States—Costs. 2. Water-supply—United States—Case studies. I. Frederick, Kenneth D. II. Gibbons, Diana C.
TD223.S29 1985   333.91'11'0973   85-20999
ISBN 0-915707-21-7

Resources
FOR THE FUTURE

1616 P Street, N.W., Washington, D.C. 20036

DIRECTORS

M. Gordon Wolman, *Chairman*

Charles E. Bishop	Franklin A. Lindsay
Anne P. Carter	Vincent E. McKelvey
Emery N. Castle	Richard W. Manderbach
William T. Creson	Laurence I. Moss
Henry L. Diamond	Leopoldo Solís
James R. Ellis	Carl H. Stoltenberg
Jerry D. Geist	Russell E. Train
John H. Gibbons	Barbara S. Uehling
Bohdan Hawrylyshyn	Robert M. White
Thomas J. Klutznick	Franklin H. Williams

HONORARY DIRECTORS

Horace M. Albright	Edward S. Mason
Hugh L. Keenleyside	William S. Paley

John W Vanderwilt

OFFICERS

Emery N. Castle, *President*
John F. Ahearne, *Vice President*
Edward F. Hand, *Secretary-Treasurer*

Resources for the Future is a nonprofit organization for research and education in the development, conservation, and use of natural resources, including the quality of the environment. It was established in 1952 with the cooperation of the Ford Foundation. Grants for research are accepted from government and private sources only on the condition that RFF shall be solely responsible for the conduct of the research and free to make its results available to the public. Most of the work of Resources for the Future is carried out by its resident staff; part is supported by grants to universities and other nonprofit organizations. Unless otherwise stated, interpretations and conclusions in RFF publications are those of the authors; the organization takes responsibility for the selection of significant subjects for study, the competence of the researchers, and their freedom of inquiry.

The book is a product of the Renewable Resources Division of Resources for the Future, Kenneth D. Frederick, director. It was designed by Elsa B. Williams and Martha Bari and edited by F. R. Ruskin. The index was prepared by Max Franke.

Contents

PREFACE xi

1 OVERVIEW *by Kenneth D. Frederick* 1

    Limitations of Traditional Solutions for Current Water Problems 2
    About This Book 4
    Emerging Issues 6
    Case Studies 12
    Conclusions 19

2 COMPETITION BETWEEN IRRIGATION AND HYDROPOWER IN THE PACIFIC NORTHWEST *by Walter R. Butcher and Philip R. Wandschneider, with Norman K. Whittlesey* 25

    Water Supply and Use in the Columbia River Basin 26
    Water and Energy Tradeoff 31
    Actors and Institutions 39
    The Swan Falls Case in Idaho: Intrastate Irrigation-Hydropower Conflict 44
    The Columbia Basin Irrigation Project: State Versus Federal Control of Water 52
    Improving Water Allocation 57
    Conclusions 63

3   WATER SCARCITY AND GAINS FROM TRADE IN
    KERN COUNTY, CALIFORNIA by H. J. Vaux, Jr.   67

    Kern County Water History   68
    Institutions Governing Water Use and Supply   75
    Gains from Trade   89
    Barriers to Trade   95
    Conclusions   99

4   SATISFYING SOUTHERN CALIFORNIA'S THIRST FOR
    WATER: EFFICIENT ALTERNATIVES by Richard W.
    Wahl and Robert K. Davis   102

    Existing Sources of Supply   105
    Alternatives for Balancing Supply and Demand   114
    Prospects for Economically Efficient Alternatives   125

5   COSTS OF WATER MANAGEMENT INSTITUTIONS:
    THE CASE OF SOUTHEASTERN VIRGINIA by Leonard
    Shabman and William E. Cox   134

    The Southeastern Virginia Water Situation   135
    The Lake Gaston Pipeline Versus Local Supply: A Cost
        Comparison   152
    Approaches to Institutional Reform   161
    Conclusion   168

6   INNOVATIONS IN WATER MANAGEMENT: LESSONS
    FROM THE COLORADO-BIG THOMPSON PROJECT
    AND NORTHERN COLORADO WATER
    CONSERVANCY DISTRICT by Charles W. Howe, Dennis
    R. Schurmeier, and William D. Shaw, Jr.   171

    Evolution of the Project   172
    NCWCD Water Markets   183
    Efficiency of the NCWCD Water Market
        Arrangements   195
    Conclusions and Recommendations   198

INDEX   201

TABLES

2-1. Water Supply and Streamflow in the Columbia and Snake Rivers 28
2-2. Major Uses of Water in the Pacific Northwest Region, 1980 29
2-3. Developed Head and Generating Capability at Hydroelectric Power Structures on the Snake and Columbia Rivers, 1960 and 1980 33
2-4. Opportunity Cost of Diverting Water from Hydropower Production, Selected Pacific Northwest Areas, 1960 and 1980 35

3-1. Acreage and Gross Value of Irrigated Crops in Kern County, 1980 72
3-2. Sources and Uses of Water in Kern County, 1980 72
3-3. Kern County Water Districts Listed by Major Source of Surface Supply 79
3-4. Average Retail Costs for Irrigation Water per Acre-Foot in Selected Kern County Water Districts, 1981 88
3-5. Estimated Annual Water Supply Deficiencies in Kern County Districts, 1980 91
3-6. Estimated Annual Gains from Interregional Trade Between Kern County and Northern Agricultural Region 93
3-7. Surface Water Prices of West Side and East Side Water Districts in Kern County, 1981 94

4-1. Metropolitan Water District Estimates of Service Area Supply and Demand 106
4-2. Priorities for Colorado River Water in California (Seven Party Agreement) 110
4-3. Use of Colorado River Water 111
4-4. Potential Future Yields from State Water Project and Central Valley Project 114
4-5. Estimated Water Losses and Costs of Conservation Investments in Imperial Irrigation District 121
4-6. Total Cost of Conserving Water in the Imperial Irrigation District and Delivery to the Metropolitan Water District 123
4-7. Summary of Alternatives for Balancing Supply and Demand in Metropolitan Water Districts 126

5-1. Projections of Water Use for Southeastern Virginia   138
5-2. 1980 and 2030 Projections of Per Capita Water Use   139
5-3. Surface-Water Yields and Well Capacity for Southeastern Virginia Water Systems   140
5-4. Surface-Water Deficit for Southeastern Virginia, 1980 to 2030   152
5-5. Water Surplus or Deficit After Implementation of Surface-Water Development Activities   153

6-1. Estimated Mean Annual Virgin Flows at Lee Ferry, Arizona   176
6-2. Some Price Statistics on NCWCD Allotments   188
6-3. Comparison of Wet Year with Dry Year Rentals   190
6-4. Rental Price Summary for C-BT Water   190

FIGURES

2-1. Columbia-Snake drainage system   27
2-2. Columbia-Snake River basin: hydropower projects and primary future irrigation areas   32

3-1. Kern County, California   68
3-2. Streams and conveyance facilities serving Kern County   69
3-3. Irrigation districts of Kern County   78
3-4. Major water conveyance facilities in Kern County   81

4-1. California developments on the Colorado River   103
4-2. Central Valley Project and State Water Project   104
4-3. Water supply from conservation investments in the Imperial Irrigation District   122

5-1. The study area   136

6-1. The Northern Colorado Water Conservancy District and the Colorado-Big Thompson Project   173
6-2. Transmountain diversion features of the Colorado-Big Thompson Project   174
6-3. Comparison of NCWCD allotment price with ditch company share prices, 1975–1983   193

Preface

The United States as a whole is richly endowed with water resources. Enormous quantities of fresh water stored in the nation's ground and surface water reservoirs are supplemented by precipitation averaging nearly 30 inches per year. The combination of abundant natural resources and government subsidies for water development projects led most Americans to take low-cost, high-quality water supplies for granted. Even in the nation's arid regions, inexpensive water has generally been available to support a rapid and water-intensive growth. In the last decade, however, this complacency about water has waned as the adequacy of the nation's supplies has emerged as a principal resource concern.

Traditionally, water adequacy has been viewed almost exclusively in terms of meeting "offstream" water uses such as those for domestic and commercial purposes, irrigation, industry, and mining; concerns about shortfalls are translated into support for dams, reservoirs, wells, and canals to increase offstream water supplies, often at the expense of "instream" uses such as those for fish and wildlife habitat, outdoor recreation, and hydropower production. The Second National Water Assessment published in 1978 provided the first nationwide examination of instream flow conditions and requirements. By incorporating an allowance for instream uses into its water-supply adequacy analysis, the assessment helps identify those watersheds where water is scarce and where there is strong competition between instream and offstream water uses. The assessment fueled expectations that water shortages will become more widespread and frequent. But the assessment provided little insight into how critical the problems are, and even less about how they should be resolved. These shortcomings are an inevitable result of estimating how much water is required as a matter apart from its cost and worth, what the demand for it is, how the present institutions allocate its uses, and whether these institutions provide incentives to conserve.

In 1983 Resources for the Future undertook a project designed to help fill some of the gaps left by the second assessment. The principal products of that effort will be two forthcoming RFF works—*The Economic Value of Water* by Diana C. Gibbons and an edited volume, *Scarce Water and Institutional Change,* of which this preprint is the first chapter.

The Gibbons study provides a framework for understanding water values and summarizes a large literature on the value and demand for water in various uses and regions. By focusing attention on relative water values and the sensitivity of the quantities demanded to its price, this study offers strong evidence about the shortcomings of the tradition that assumes projected offstream water uses are to be met regardless of costs or effect on instream flows.

The edited volume uses case studies prepared by some of the nation's leading water resource experts to examine alternative ways for meeting long-term water needs in five areas of the country facing a variety of water problems. The overview chapter for this volume highlights the factors underlying the nation's water supply problems and summarizes the lessons to be learned from the case studies. In so doing, the overview stands on its own. The full volume is scheduled for publication in early 1986. I believe that advance publication of this chapter will be useful as a brief analysis of an important current resource issue.

Many people and organizations helped make the volume possible. The efforts and patience of the authors of the case studies were essential to the enterprise, and a major share of the credit for any success it achieves is theirs. Diana Gibbons provided detailed and insightful comments on early drafts of all the chapters. An advisory panel consisting of Robert K. Davis, Irene Murphy, Edgar H. Nelson, Kenneth I. Rubin, Theodore Schad, John Schefter, and Gary Taylor helped get the project off the ground with suggestions for interesting cases and possible authors. Among my colleagues at Resources for the Future, John Ahearne and Sally Skillings reviewed the entire text along with several anonymous reviewers, and Allen Kneese commented on chapter 2. Maybelle Frashure did the typing and Elsa Williams designed the book. The editing of F. R. Ruskin made the entire volume more readable. While the final product has benefited from these contributions, the authors bear responsibility for their chapters.

Major funding for this project was provided by a generous grant from the William H. Donner Foundation. Their support as well as that of Resources for the Future and the General Service Foundation are gratefully acknowledged.

August 1985   Kenneth D. Frederick

1
Overview

Kenneth D. Frederick

In 1980 and 1981, when large areas of the United States were parched by drought, alarmist reports on the state of the nation's water supplies received wide coverage. The cover of *Newsweek* for February 23, 1981, showed the United States covered with parched earth and asked, "Are we running out of water?" In *U.S. News and World Report,* June 29, 1981, the title of a feature article asked a similar question, while a subheading responded that "Nearly every part of the U.S. faces serious water troubles—either lack of supply or doubtful purity. Experts warn that time for remedies is rapidly running out." *Science 81* ran a series of articles on local water issues; the May issue contained an article entitled "The Day New York Runs Out of Water." Numerous other articles, television programs, and books (for example, Fred Powledge's *Water: The Nature, Uses, and Future of Our Most Precious and Abused Resource* and William Ashworth's *Nor Any Drop to Drink*) reinforced the image of an imminent water crisis. Water was at the forefront of resource concerns of the 1980s.

Then, as more normal patterns of precipitation returned, the crisis atmosphere receded and the urgency for action waned. Little, if anything, has been done to enable most regions to prevent or cope more effectively with future shortages of water. Although no one knows for certain when the next drought will develop, we do know that it will come. Furthermore, important longer-term problems remain unresolved. Groundwater mining, contamination of water supplies, deteriorating water supply facilities, and rising costs accompany the growing demands on water supplies. High-quality water is becoming increasingly scarce relative to the demands for it, but there is little agreement as to what should be done to prevent future problems.

Limitations of Traditional Solutions for Current Water Problems

All levels of government share the responsibility for ensuring the availability of adequate water supplies. Traditionally, water resource development has focused on managing supplies to meet offstream demands. Withdrawals have been projected to grow roughly in step with population and economic growth, and projected levels of water use have acquired the status of requirements, of virtual necessities to be provided regardless of cost. Despite evidence to the contrary, the quantity of water demanded by offstream users has generally been assumed to be insensitive to the costs. Consequently, public concern about water adequacy is translated into support for large, capital-intensive projects, and planners tend to focus on structural solutions. Dams, reservoirs, wells, and other water supply infrastructures are designed to meet projected-use levels for all but the most severe droughts. Users are expected to have to adjust to water shortages on only rare occasions.

Drought in various regions of the country from 1976 to 1977 and 1980 to 1981 produced the expected surge in the demand for investment in water supply facilities. But such investment encountered more than the customary resistance and scrutiny, and many of the proposed water supply projects have yet to move beyond the planning stage. High interest rates and budgetary concerns have made federal funding more difficult to obtain. More importantly, the costs of developing new water supplies now often appear high in comparison with the benefits to be derived.

The least-expensive sources of supply were developed first. Initially, this meant simply diverting water from neighboring streams or using groundwater. But only a small part of a stream's average annual flow is available on a continuing basis, because of the seasonal nature of most surface flows. Reliable offstream water supplies can be increased only at the cost of constructing facilities to compensate for this irregularity.

In the early stages of development, costs were contained by taking advantage of the best natural reservoir and dam sites or by utilizing easily accessible groundwater. But three factors make increasing costs for new offstream supplies inevitable. First, because the best sites are developed first, the costs of additional units of storage rise as the capacity of the system increases. Second, as storage capacity increases, a stream's safe yield (the quantity of water that can be supplied with some high degree of probability) increases, but only at a diminishing rate. Finally, because marginal values of instream flows rise as the resource becomes scarcer, the opportunity costs of storing and diverting water rise as instream flows diminish.

The traditional, structural approach to preventing and solving water problems may have approximated an efficient strategy when the direct costs of providing reliable supplies were low and streamflows were sufficient to meet all demands. For instance, when extraction of water for one use does not affect its availability for other uses, the opportunity costs of the withdrawal are effectively zero. In the same way, when the assimilative capacity of water bodies is not exceeded, waste disposal is not competitive with other water uses. Moreover, when large quantities of water can be developed at relatively low cost, it may not be unreasonable to assume that the value of developing new supplies exceeds the costs. Unfortunately, such conditions no longer characterize the water situation in most of the nation. The total costs of increasing offstream supplies are generally high and likely to exceed the value of many water uses.

Nontraditional means of increasing water supplies, such as desalination, transporting icebergs, and weather modification, offer no panacea. Desalination costs are generally higher than those for most traditional water projects; towing icebergs to arid regions poses technical and legal as well as cost problems; and although winter cloud seeding is a cost-effective means of increasing water supplies under some conditions, its application is limited both geographically and by legal considerations. Although use of desalination and weather modification will continue to grow on a limited basis, none of these measures is likely to become the source of large quantities of water in the foreseeable future.

The limitations of supply-side solutions to the nation's water problems mean that greater attention must be given to demand-side management to resolve long-term problems. This important lesson has not yet been fully learned. Measures to limit water use tend to be viewed as short-term responses, to be used only when crisis looms and supplies appear insufficient to satisfy demand at prevailing prices. Commonly, water is considered too essential a resource or too insensitive to price for its use to be left to the impersonal forces of free markets during shortages. Direct controls or patriotic appeals to conserve are often resorted to under such conditions. And despite the high costs of moving water long distances out of natural channels and the fact that few, if any, areas are now willing to give up supplies without compensation, the hope persists that imports will replenish supplies in regions facing severe long-term problems of water supply.

About This Book

Because water problems and their feasible solutions tend to be local or regional in nature, case studies form the core of this volume. The first four studies focus on areas experiencing problems with projected water needs or with allocating scarce supplies among competing uses. The studies of the Pacific Northwest, Kern County in California, the coastal area of southern California, and southeastern Virginia examine the water situations in these locations and assess alternative means of meeting their long-term water needs. The final study, which examines the development and operation of the Colorado-Big Thompson Project in northeastern Colorado, is a detailed look at an innovative institutional arrangement, one that has facilitated the reallocation of water in response to changing conditions.

Individually, the case studies provide interesting, well-documented analyses of specific water issues and institutions; collectively, they offer insights into the nature and causes of water problems and their solutions that can be helpful in other areas as well. Topics addressed in the case studies that have broad relevance for water supply management include:

- the role of institutional factors, especially water laws and administrative arrangements, in creating and solving a region's water problems;
- the factors that promote or inhibit flexibility for responding to changing water supply and demand;
- the potential role of markets and prices for allocating scarce water supplies and creating incentives to conserve water; and
- institutions and conditions that encourage water marketing.

The case studies are written by economists and reflect the economist's traditional concern with economic efficiency. The papers do not, however, assume that efficiency should be the sole criterion for managing scarce water resources or that greater efficiency in the allocation and use of water is easily attained. Institutional barriers and equity implications of proposed water-management alternatives receive careful consideration. On the other hand, there is no attempt to present a comprehensive examination of all water-management issues in any of the case studies. For Kern County, southern California, and Virginia Beach, the focus is the quantity of water available for offstream use; for the Pacific Northwest, the focus is the allocation of water between irrigation and hydropower; and for northeast Colorado, the focus is the institutional mechanism for transferring water among users.

The case studies emphasize the quantitative dimensions of meeting long-term demands and resolving conflicts among water uses. However, water quality may be an important consideration in developing an efficient strategy for meeting water needs and in assessing the adequacy and the true cost of alternative supplies. Furthermore, efficient water management may have to account for the impacts a given use has on the quality—and therefore the value of—return flows. The southern California study exemplifies a situation where consumers are not indifferent about alternative sources of supply. In comparing the total delivered costs of alternative supplies, a penalty in excess of $100 per acre-foot is used to adjust for the higher salinity levels of Colorado River water. Water quality is not a central issue in the other case studies. Nevertheless, the authors do call attention to some water quality issues that any comprehensive study of the water situations in the various areas would have to address.

This volume is intended for a broad audience; anyone seeking a better understanding of the nation's water supply issues and their solutions should find the book both useful and easy to understand. Economic jargon is kept to a minimum for the general reader, but the water resource specialist will find new data and analysis in the case studies that should be of interest.

The nature, magnitude, and urgency of the problems differ greatly among the regions. For instance, the water supply problems of Virginia Beach pale in comparison with those of southern California. Virginia Beach lies within a water resource subregion where offstream consumption is only 2 percent of average streamflow, whereas offstream consumption in southern California exceeds average annual streamflow (U.S. Water Resources Council, 1978b). Further, southern California is losing some of the imported supplies that meet current water needs. Nevertheless, both regions share the concern that future water shortages could produce major hardships and inhibit development. The water problems of two such vastly different areas bear out the universality of the central themes of this volume—the importance of institutional factors in creating water problems and the potential of institutional change as a means of resolving them.

This introductory chapter is a brief overview of some of the principal factors that influence how the nation uses and abuses its waters; of important issues in need of resolution; and of the role of water markets. The overview is followed first by summaries of the issues addressed in the case studies and then by comments on the broader lessons to be learned about the nature of our water problems and ways to resolve them.

Emerging Issues

Water Uses

The United States is a water-rich nation. Annual precipitation over the forty-eight contiguous states averages nearly 30 inches per year, or 4,200 billion gallons per day (bgd). In addition, large quantities of freshwater are stored in surface water and groundwater reservoirs.

With government policies encouraging and often subsidizing the development of supplies, the availability of high-quality water at very low prices has come to be taken for granted by most citizens. It is not surprising, therefore, that the United States uses large quantities of water. As of 1980, total freshwater withdrawals averaged 380 bgd or 1,600 gallons per person per day. Publicly supplied water (for domestic uses; for public uses such as firefighting, street washing, and municipal parks and swimming pools; and for industrial and commercial uses) averaged 34 bgd or 183 gallons per day for each individual served (Solley and coauthors, 1983). Water use for these purposes appears low in comparison with withdrawals for irrigation and self-supplied industrial uses, which together accounted for about 90 percent of the total.

Water use per capita in the United States for household and municipal purposes is two to four times the levels in France, Germany, the United Kingdom, and Sweden (Rogers, 1983). Per capita differences in housing and yard space undoubtedly contribute to these water use differentials. But the fact that European consumers generally pay from 50 to 350 percent more for water than their U.S. counterparts is also an important factor in relative water use rates.

Relative prices also help explain why regional water use patterns within the United States tend to be inversely related to natural supplies. The seventeen western states receive only about one-fourth the rainfall per acre, yet withdrawals are nearly double and consumption is ten times greater than per capita levels in the East. The relative importance of irrigation, which accounts for 90 percent of the consumption and 74 percent of the withdrawals in the West, explains much of this regional difference. But the growth of western irrigation has depended in part on some of the lowest water prices in the nation. Even urban residents in the West generally pay less than half the price paid for water in the East, which helps explain why publicly supplied withdrawals are 45 percent higher per capita in the West (Solley and coauthors, 1983).

Increasing withdrawals from the nation's streams highlight the importance of instream water uses, for instance, for recreation, wildlife habitats, hydropower, and navigation. Instream flow requirements, measured as the minimum levels for fish and wildlife maintenance, have been estimated at about three times the nation's withdrawals (U.S. Water Resources Council, 1978a). Total water use, defined as the sum of offstream consumption and instream needs, is about 87 percent of the nation's average streamflow. Need and availability, of course, are by no means regionally matched. The competition for scarce supplies is particularly strong in the seventeen western states. In fact, total water use actually exceeds average-year streamflows in twenty-four of the West's fifty-three water resource subregions. These shortfalls in renewable supplies have resulted in extensive groundwater mining that averages more than 20 bgd in the West and contributes to rising water costs. Under conditions of moderate drought that occur, on average, every five years, streamflow in forty-eight of the fifty-three western subregions drops below mean levels of water use (U.S. Water Resources Council, 1978b).

Offstream water users have been insulated from the rising costs of new water supplies by a tendency to overlook (or at least not treat as project costs) the impacts on the quantity and quality of instream flows, by subsidies for water supply projects, and by water pricing practices. Until federal environmental legislation of the 1970s mandated greater consideration of instream values, the impacts of water projects on instream water values other than hydropower and navigation were ignored and hence not reflected in project costs. Official concern was limited largely to furnishing water for offstream uses. Irrigation, the sector most likely to be affected by higher water costs, has been the most insulated. Federal subsidies for Bureau of Reclamation irrigation projects often exceed 90 percent of construction costs (Frederick with Hanson, 1982). And for most other consumers, average cost pricing masks the rising costs of new water supplies by averaging the high costs of additional water sources with the lower costs of established supplies.

The growing gap between the costs of developing new supplies for offstream use and the prices users pay for the water encourages use beyond economically efficient levels and increases the demand for new water supply projects. As noted, however, forces are emerging to counter this push for more water projects, as projects are subjected to closer scrutiny by those concerned with budget deficits, environmental issues, and competing uses of the water. Accordingly, greater attention is now being given to the institutions that allocate limited water supplies among competing uses.

Water Law

Water is not treated like any other resource. The states reserve the power and prerogative to establish the institutions for allocating all the waters within their boundaries not encumbered by federal law or interstate compact. The states grant water use rights based on either a common law doctrine that calls for all users to cut back in time of shortage or a system in which the earliest users have the most senior rights. In almost all cases, water is treated as a free commodity; charges are not made for extracting water from surface or groundwater sources, but only for the costs of moving the water. The rights to the water, however, are often constrained in ways that limit or at least raise doubts about the legality of transfers to other uses and users.

The earliest state laws controlling surface waters were based on the common law doctrine of riparian rights, which grants the owner of land adjacent to a water body the right to use the water. Riparian rights are inseparable from the land and are constrained to uses that are "reasonable" and do not unduly inconvenience other riparian owners. There is no priority of use, so all riparian owners must share in curtailing use in time of shortage. Riparian doctrine still underlies the water codes of almost all the relatively water-abundant eastern states.

In regions where streams are less numerous and their flows smaller and less reliable, extensive development required diverting water beyond riparian lands with greater assurance of availability. Consequently, the seventeen western states adopted the doctrine of prior appropriation as the basis of their water laws. Under this doctrine, water is allocated according to the principle of "first in time, first in right." Thus, the holders of senior water rights have priority over all subsequently acquired rights, and the full burden of any water shortage is borne by the holders of the most junior rights.

Appropriative rights eliminate a major obstacle to water transfers and markets by breaking the link between water and land. However, a variety of legal provisions in states using the appropriation doctrine inhibit the creation of well-defined, transferable property rights in water. Appropriative rights are acquired by diverting water from a stream and putting it to some beneficial use; the right originates only at the point of diversion and is contingent upon continued beneficial use. The acquired right can be sold or transferred, but the right is limited to use (not ownership) of the water, and there may be restrictions as to how and where it can be used. Twelve western states specify a ranked preference of use that allows preferred uses (municipal and domestic first, often followed by agriculture) to supersede water rights destined for less-preferred uses in time of scarcity. And the right to transfer water to other uses and locations is complicated by the fact that return flows are public property, freed for appropriation by downstream users.

State laws and institutions guiding the allocation and use of water have evolved over time in response to new conditions. However, as the cases in this volume indicate, changes enabling institutions to deal efficiently with water scarcity have lagged well behind need. Most of the laws and institutions remain more appropriate for an era when water was in actuality as well as in law a free resource. These provisions served their regions reasonably well when new demands could be met by developing new supplies. But their shortcomings are apparent when there is need to apportion existing supplies among alternative uses or respond to short-term shortages.

Water Markets

Markets are the customary means in our society for allocating scarce resources. Well-functioning markets direct resources to highest-value uses and provide the incentives for developing new supplies and conserving use. Markets also enable resources to flow to new uses in response to changing conditions, and they eliminate shortages and surpluses through adjustments in supply and demand to price changes.

A precondition for the operation of effective markets is the existence of well-defined, transferable property rights. Riparian rights, which tie water use to riparian lands, are the most restrictive of transfers. But even appropriative rights are often severely encumbered. Restrictions on water rights comprise a powerful force to keep water flowing to established uses, where it is employed with traditional, often inefficient methods. However, incentives to conserve water are stifled when there is fear that the rights to any water savings will accrue to others. Selling a water right in some states can be risky because the sale may be taken as evidence of nonbeneficial use and thus as cause for revocation of water rights.

Most western states do permit water transfers under certain conditions so long as third-party interests are protected. These conditions may be very difficult to fulfill, however. Fees, delays, and uncertainties of public-agency approval create costs that can be large relative to the average value of water in a region, especially if litigation is involved. Thus, differences in the value of water among uses may have to be large to justify the effort and risk of selling the water rights. The often-heard phrase in the West that "water flows uphill to money" may be accurate, but the lack of market-type institutions is apt to make the priming of the pump agonizingly slow and expensive.

The view that institutional deficiencies underlie the nation's water problems has gained considerable support in recent years. But there is less agreement as to what should be done to correct these shortcomings.

Libertarians emphasize the weaknesses of centralized controls and argue that well-defined, transferable property rights are essential for establishing the markets that could improve water use. Centralization allegedly places the control of water resources in the hands of bureaucrats who lack the information and incentives for efficient water allocation. Also, government control encourages entrepreneurs to engage in nonproductive activities such as lobbying for subsidies and other special considerations. Market enthusiasts propose creation of clear, transferable property rights in water through the elimination of government-imposed restrictions on water use and ownership.

Very different conclusions are reached by those emphasizing the deficiencies of markets for allocating water. These people say that not only must there be well-defined property rights, but that individuals must face the full costs as well as the benefits of their use or exchange of the resource if markets are to work effectively.

The very nature of the resource may make it difficult, however, to fulfill either condition. First, surface waters and groundwaters that flow from one property to another are naturally common-property resources that belong to no one until they are withdrawn for use. The fugitive nature of such resources creates special problems for establishing clear property rights. For instance, groundwater resources are often situated so that one person's pumping adversely affects the quantity and costs of water available to others. Thus, the user does not bear the full costs of groundwater use unless the aquifer is placed under single ownership. When ownership is established only by extraction, private incentives cause the resource to be exploited more rapidly than is collectively desirable.

Externalities or third-party effects are a second type of market failure common to water resources. An externality results when a transaction affects someone other than the buyer and seller or when a use of the resource affects someone other than the user. In either case, the conditions required for efficient resource allocation through markets are violated. Externalities almost always result when a transfer alters either the point of diversion or return flow. Further, many of the uses to which water is put generate important externalities. For instance, most of the costs of using surface waters for waste disposal are imposed on downstream users rather than on the polluter.

Water resources produce public as well as private benefits. A public benefit cannot be successfully marketed because nonpaying individuals cannot be excluded from enjoying it. Thus, there is an incentive for users to free-ride and for private producers to underinvest in the production of such goods. Private firms and individuals, for example, underinvest in instream flows that preserve magnificent scenery and good fishing when they are unable to capture the full benefits of these goods in the marketplace. Consequently, provision of such goods is traditionally considered a government responsibility.

Indian and Federal Water Rights

A further complication in developing markets in water rights is the existence of potentially large, but so far unquantified, claims for Indian and federal reserved water rights. The basis for these claims lies in the enormous landholdings withdrawn from the public domain for Indian reservations or to achieve some other national purpose. When the lands were withdrawn, the federal government neglected to assign water rights to them. However, the Supreme Court ruled in 1908 and on subsequent occasions that when the federal government withdrew lands for any purpose, it also implicitly withdrew sufficient unappropriated water from the public domain to accomplish the purposes for which the land was withdrawn. Thus, Indian and federal reserved claims dating back to the nineteenth and early twentieth centuries have high priority in the use of western waters.

These claims create great problems for some western states. The states have generally assumed jurisdiction over all nonnavigable water in the public domain, and in the West user rights have been granted to most of the surface waters. These rights were granted with no allowance for potential Indian or federal rights that exist outside of state law and can only be fulfilled by terminating or reducing the rights of some existing users. The Reagan Administration has rejected the idea that there are federal water rights that exist independently of state water law. But even if this view is accepted by the Congress and the courts, Indian claims— which are by far the most important group of unsatisfied water rights— still remain. Consequently, in huge areas great uncertainties surround water rights with a seniority that postdates the creation of neighboring Indian reservations. Conflicts between Indian and other water users exist in at least sixty western water basins, and some of the unquantified claims are potentially large. For instance, the Navajos have threatened a court suit claiming one-third of the flow of the Colorado River. The court's recognition of tribal fishing rights in the Columbia River has led to water reallocation from power to fishing and a large, improved fishing program funded by power revenues. The uncertainties surrounding unresolved Indian claims discourage development and hinder the assignment of clearly defined, transferable property rights that might encourage improved water use.

Case Studies

Columbia River Basin

The first case study focuses on a relatively water-rich area, the Columbia River basin. Walter Butcher, Philip Wandschneider, and Norman Whittlesey examine how technical and economic factors have brought major changes in the relative instream and offstream values of water in the Pacific Northwest, and they assess the adequacy of existing institutions for dealing with these changes. The most striking change is reflected in the value of the energy forgone when water is diverted for irrigation; the replacement cost of power loss because of depletion of an acre-foot of water in southeast Idaho increased from $1.70 in 1960 to $64 in 1980.

These relatively high instream values are not well reflected in water allocations because rights for instream uses have traditionally been subordinate to those for offstream use. Irrigators have enjoyed the right of free access to water to develop new irrigated lands even when the withdrawals were detrimental to established instream uses. With the recent rise in the value of water for hydropower, power interests have sought to strengthen their own rights to instream flows. The struggle between irrigation and power uses of Columbia basin water is illustrated with two examples—the Swan Falls case, involving intrastate conflicts, and the Columbia Basin Project, involving federal-state issues.

The Swan Falls case emerged in a suit by the Idaho Power Company against irrigators and the State Department of Water Resources for illegally "taking" water from the Snake River, a major tributary of the Columbia, that had been granted to the company for electricity production. Water rights for hydropower production at Swan Falls have an early priority claim to water in the Snake River, but it was widely believed that hydropower rights in Idaho were subject to preemption by irrigation and other diversionary uses. By the mid-1970s the power company was faced with declining hydropower output at existing dams because of declining river flows, just as the company faced the need to produce much higher cost thermal power to meet growing energy demands. The prospect that by the year 2000 an additional 1.7 million acre-feet per year would be depleted from Snake River flows if state plans for irrigation development were fulfilled, led the power company's unhappy customers to complain to the Idaho Public Utility Commission that the company was failing to defend its water rights at Swan Falls. The 1982 decision of the Idaho Supreme Court upholding the power company's right to a flow of 8,400 cubic feet per second (cfs) at Swan Falls Dam had major implications for the allocation of water and, therefore, wealth in the region. Hope for new irrigation and even the rights of existing irrigators were placed in jeopardy by the court's decision. The Idaho legislature moved to protect existing irrigation, and attempts are underway to reestablish the opportunity for more irrigation in the Snake River plain. In the absence of legislation clarifying water rights, many cases like Swan Falls may be necessary to define and specify the rights of instream and offstream users.

The proposed 500,000-acre expansion of the Columbia Basin Irrigation Project is another example of water conflict in the basin. The project could divert as much as 2.5 million acre-feet per year of water from the Columbia River above the Grand Coulee and ten other hydropower dams. The water would irrigate already productive wheat lands at an investment cost in excess of $4,000 per acre. In addition, replacement of lost hydropower would increase the costs of supplying the region's electricity by more than $200 per year for each acre irrigated.

Irrigators cannot afford to pay more than a small fraction of these costs, but they anticipate generous subsidies from the state of Washington, the Bonneville Power Administration, and the federal Treasury to defray the capital costs. Furthermore, proponents of the project argue that the water rights granted the Bureau of Reclamation for completion of the Columbia Basin Project have priority over hydropower users, and therefore, the project bears no responsibility for downstream effects. Accordingly, prospective irrigators favor the project that otherwise would make a very poor investment.

The power interests, on the other hand, may not readily acquiesce in projects that divert the water they have been using for some years. Butcher and his coauthors point out that there are uncertainties surrounding the project's water rights and that the subordination of downstream hydropower rights has not been tested. Recent loss of water and substantial rights for fishery restoration may make power interests even less sympathetic to new irrigation. Although the fate of the Columbia Basin Irrigation Project is in doubt, the analysis demonstrates the tenuous nature of the rights hydropower users have to water in the Columbia basin and that existing institutional arrangements permit irrigation to expand even when expansion will surely reduce the net economic benefits the region derives from its water resources.

The existing institutional structure for allocating Columbia River water is a legacy of a time when instream flows could reasonably be assigned zero marginal value. The situation has changed dramatically in the last two decades and a major struggle for water has emerged with enormous equity implications. As the authors point out, however, the important issue for the economic health of the region is not who initially owns the water but, rather, ensuring that the water can be reallocated, when necessary, to new high-value uses. Thus, rights to the water need to be unambiguous and unattenuated and some mechanism established to ensure that water can be transferred among uses in response to changing economic opportunities.

California

Chapters 3 and 4 present case studies of two areas in California, Kern County and the Los Angeles-San Diego metropolitan area, which are encountering problems both in maintaining existing water sources and in increasing supplies. Both areas had counted on the State Water Project (SWP) to augment their future water supplies.

Authorization for the ambitious project stems from 1959 legislation that authorized construction of a major aqueduct and storage system to move large quantities of water from the well-watered northern parts of the state to the arid central and southern parts. To ensure a market for the water and to cover the costs, the state made contractual commitments for the eventual delivery of 4.23 million acre-feet of water annually. Facilities capable of supplying just over half of this water were in place by 1980 when legislation was passed authorizing projects that would add 1.9 million acre-feet to the capacity of the SWP at a cost in excess of $2.9 billion in 1980 dollars.

The most controversial part of the SWP plan was the Peripheral Canal, a 43-mile aqueduct proposed to carry water from the Sacramento River, around the Sacramento-San Joaquin River delta, to the California Aqueduct from whence it could be shipped south. In 1982, a statewide referendum overturned the 1980 legislation, leaving considerable doubt as to how the state could meet its water supply obligations.

With existing facilities, the SWP has a dependable yield of about 2.3 million acre-feet annually. But this yield will drop by 0.5–0.7 million acre-feet as water use increases in the basins of origin and in the federal Central Valley Project. In the absence of new storage, contracts for SWP water will exceed installed capacity by the year 2000 by as much as 1.4–1.6 million acre-feet in a dry year. Concerns about the state's ability to deliver on these contracts have led several areas to explore alternative approaches to meeting water needs. The Kern County Water Agency and the Metropolitan Water District (MWD), which supplies water to six southern California counties, are two agencies with such concerns.

Kern County, California, has 944,000 acres under irrigation, producing crops valued at well over $1 billion annually. With an average annual rainfall of only 5–8 inches, virtually none of which falls during the long, hot growing season, the region's economy depends on abundant supplies of imported water. In chapter 3, Henry Vaux, Jr., examines emerging water problems in the county as well as the obstacles to and potential gains from water transfers. The supply of about 1 of every 8 gallons of water now used by the county is insecure. Current groundwater use is depleting stocks by an average of 370,000 acre-feet annually. Further, 200,000 acre-feet of SWP water now going to Kern County will be lost to southern California when the Central Arizona Project opens and California is forced to curtail diversions from the Colorado River. Although Kern County's contracts for SWP water rise to 1.25 million acre-feet in 1990, it is unlikely the state will be able to deliver much more than 900,000 acre-feet per year before the year 2000, and the prospects after that date are uncertain. Consequently, the county's continued prosperity, which is intimately linked to irrigated agriculture, depends on how well it can adjust to the increasing scarcity of water.

Vaux echoes a familiar theme of this volume as he makes a persuasive case for introducing some means of facilitating water trades among the county's various water districts and with other areas of the state. His recommendations are supported by model results suggesting that sales of water to Kern County from agricultural water users to the north could benefit both regions. Vaux also documents how administrative review of proposed transfers results in high transactions costs, which, along with existing restrictions on water rights, diminish the benefits and even the opportunities for transfers. The author is not optimistic about the prospects for facilitating trading in private water rights; he concludes that it is likely to be too costly and difficult to eliminate the barriers to trade. Water districts, on the other hand, could be the focus of active water markets if a few restrictions on their activities (that serve to inhibit water sales but fail to protect third-party interests) were eliminated and if incentives to trade were provided by allowing the districts to make a profit.

The spectacular growth of southern California's coastal region has depended on three major sources of imported water—the 240-mile Los Angeles Aqueduct, which draws water from Owens Valley and Mono Lake basin; the 200-mile Colorado River Aqueduct, which brings water from below Hoover Dam; and the 500-mile California Aqueduct, which carries water from the Sacramento-San Joaquin delta. Increased competition for these waters and lawsuits to enforce environmental legislation threaten to curtail the region's existing supplies and hamper their search for new ones. From 20 to 40 percent of the water currently taken each year from Owens Valley and Mono Lake could be lost, depending on the outcome of lawsuits now in the courts. The MWD's withdrawals from the Colorado River, which have ranged from 0.711 to 1.285 maf per year over the past decade, will drop to about 0.55 maf per year when Arizona starts utilizing its full entitlement to the river. And although the MWD can claim an additional 200,000 acre-feet of SWP water that it has not needed because of the availability of surplus Colorado River water, as already noted, further increases from the SWP are not likely for some time. By the year 2000, MWD estimates suggest that available supplies even in a year of average rainfall will fall 140,000 acre-feet short of projected water use.

The study by Richard Wahl and Robert Davis examines the alternatives for meeting these shortfalls. The least-expensive and most-efficient approach involves demand management. Introduction of marginal cost pricing in southern California would reduce water use sufficiently to obviate the need for adding to supplies before the year 2000. This alternative, however, would require at least a 77-percent increase in water rates, an adjustment that is not contemplated by the area's officials. Nevertheless, higher water rates will be unavoidable as the region adds higher-cost sources of supply to its system and disposal costs rise. These rates will dampen the increase in demand.

MWD officials continue to take the position that additional SWP water is essential to meeting their long-term water needs. But the least costly addition, the Peripheral Canal, already has been rejected by the voters. The costs of providing water to the MWD through the other facilities under consideration by the state are estimated to be about 50–100 percent higher than the $216 per acre-foot cost estimated for the Peripheral Canal.

Lower-cost water is potentially available from the federal Central Valley Project (CVP), which has an expected surplus of nearly 1.5 million acre-feet of water not under long-term contract. California has offered to purchase this water for delivery to agencies that have contracted for SWP water, but the federal government has not acted on its proposal. Even if an agreement were to be reached, its duration might be limited by the need to recognize the interests of agricultural water users.

Another alternative that Wahl and Davis examine in some detail is for the MWD to buy water saved through conservation investments in the Imperial Irrigation District (IID). Up to 0.4 million acre-feet could be delivered to the MWD at marginal costs ranging from $200 to $300 per acre-foot with no reduction in IID irrigation. Given the opportunity, IID farmers would probably be willing to sell even more water to the MWD at prices attractive to both parties. Although there are institutional barriers to such transfers, support is growing within California for transfers and may soon be sufficient to remove the obstacles.

Southeastern Virginia

Southeastern Virginia is an unlikely candidate for problems in meeting long-term water supplies. The southeastern Virginia water supply planning area of Virginia Beach, Norfolk, Chesapeake, Portsmouth, Suffolk, and Isle of Wight has ample supplies for all its needs. Virginia Beach alone, however, is not fully self-sufficient and requires the cooperation of neighboring jurisdictions to meet the anticipated water needs of its rapidly growing population. Such cooperation has not always been forthcoming, and many Virginia Beach residents foresee a water crisis if they are unable to import water from outside the planning area. Fears of an impending crisis have been fueled by the local press. For example, an editorial in *The Virginian-Pilot* (August 4, 1984) denouncing the tactics employed by North Carolina legislators to delay construction of a pipeline proposed by Virginia Beach laments, "Meanwhile, while these legislative skirmishes are unfolding—and parallel legal battles over the pipeline are creeping through the courts—the water clock keeps ticking relentlessly for Virginia Beach and the rest of South Hampton Roads. Projections show that severe water shortages are possible by the early 1990s. Will the pipeline be a reality before the faucets run dry?"

Analysis of Virginia Beach's water situation by Leonard Shabman and William Cox in chapter 5 suggests considerable exaggeration by the press of Virginia Beach's plight. Their analysis indicates that institutional shortcomings and conflicts among neighboring political jurisdictions are at the root of the problem. Meanwhile, the concern has led Virginia Beach to propose construction of an 85-mile, $172-million pipeline to Lake Gaston to fill a projected gap between its supply and demand for water. In comparison with a conjunctive-use alternative, which the authors argue would be possible under a more-favorable institutional setting, the proposed pipeline would add about $60 million to the costs of Virginia Beach's water.

The conjunctive-use alternative would require the creation of a fully integrated regional water system. Water supplies within the system could be physically integrated by (1) using existing groundwater capacity as a supplemental source in time of drought, (2) constructing a 15-mile pipeline to complete the interconnection of reservoirs within the planning area, (3) expanding the use of surface water from the Blackwater-Nottoway River System, and (4) developing a drought-management plan to expand the joint yield of the separate reservoir, river, and well systems. However, the state's failure to clarify groundwater rights and to develop a clear policy on groundwater use make this a risky alternative.

Despite the possibility of a lower-cost solution, Shabman and Cox conclude that under the existing institutional setting Virginia Beach is justified in proposing the Lake Gaston pipeline. But this interbasin import alternative faces its own institutional obstacles: North Carolina is contesting Virginia Beach's rights to withdraw water from Lake Gaston. Although the authors give North Carolina's claims little chance of preventing the exports from Lake Gaston, their legal protests contribute to Virginia Beach's anxieties over securing additional water and, undoubtedly, will add to the cost and time required to complete the project.

Colorado

A common theme of the first four studies is the need for increasing water transferability among uses and jurisdictions. The authors all speculate about the types of institutional changes required to bring this about within their study areas. The final paper by Charles Howe, Dennis Schurmeier, and William Shaw, Jr., describes and evaluates an innovative institutional arrangement that facilitates the kinds of transfers that seem to be needed in the other areas.

The Bureau of Reclamation's Colorado-Big Thompson (C-BT) Project brings an average of 230,000 acre-feet of water annually from the western slopes of the Rocky Mountains to northeastern Colorado. Several innovative features of this federal project make it possible to actively trade the rights to proportional shares of this water. From the perspective of establishing water markets, the most important difference between this and other Bureau projects is that the United States retains ownership of all return flows. This enables transfers of project water without worry about downstream, third-party liabilities.

This feature is an exception to western water law, where return flows are treated as part of the stream and subject to appropriation by downstream users. The exception, which was justified by the fact that the project provided new water to the eastern slope, does not eliminate third-party impacts of water transfers, but it does eliminate the buyer's and seller's responsibility for them. Howe and his coauthors argue that within the setting of the Colorado-Big Thompson Project the advantages of being able to transfer water easily and rapidly from agriculture to growing municipalities and industries more than offset the inefficiencies that result from ignoring third-party effects. Although they are generally laudatory in their evaluation of the market established in C-BT water, the authors have recommendations for improving its operation.

A broader implication of the study is that the institutions for facilitating water transfers need to be adapted to individual circumstances. The paper offers guidance as to when proportional or priority allocation rules are most appropriate and when third-party impacts can or cannot be advantageously ignored.

Conclusions

Five case studies cannot provide the basis for sweeping generalizations regarding the nation's water problems. Nevertheless, there are lessons to be learned from these studies with relevance for understanding and resolving a broader set of water-related problems. Although the wide range of institutional arrangements, problems, and water endowments covered by the case studies may obscure points of similarity among them, it may also broaden the applicability of these common elements.

In spite of their vastly different water supply conditions, the four water-problem regions (the Columbia River basin, Kern County, southern California, and Virginia Beach) examined in this volume share a concern that water shortages might stifle growth and impose hardships on important parts of their communities. In varying degrees, water is a scarce resource for each of these areas, and the easily accessible, low-cost sources of new supplies have either already been exhausted or their development is obstructed by institutional factors.

Scarcity is characteristic of most resources and is something societies must constantly deal with. Fortunately, in most cases they manage to do so without causing undue alarm. The case studies suggest why water is often the exception. But contrary to the view popularized in the media, it is not because these areas are running out of water.

The basis for the fear of water shortages lies in several related factors. First, a willingness to incur higher costs is no guarantee that additional water will be forthcoming; adding to supplies is likely to require either the approval of recalcitrant neighbors or diverting water from valued or entrenched alternative uses. Competition for water is often resolved only after costly and protracted legal battles, begun with considerable uncertainty about the outcome. Second, the existence of supplies going to low-value uses or being used inefficiently does not guarantee they can be diverted to more urgent needs in time of shortage. In the areas where demand places the greatest pressures on existing supplies, large quantities of water can either go to relatively low-value uses or, given the proper incentives, can be conserved with more-efficient methods of distribution and application. Institutional factors, however, often restrict transfers among uses and limit incentives for efficient use. And third, when water rights are clouded in uncertainty, a region's development is likely to suffer. The higher risks or costs associated with attaining water supplies discourage new investments. Even the rights of established water users have been challenged in the courts in recent years. Thus, it is not scarcity per se that justifies the fear of costly water shortages in the study areas. The roots of the concerns of future shortages lie in the laws, administrative practices, and other institutions that create uncertainty over water rights, pose obstacles to developing new supplies or reallocating existing supplies to new uses, and provide little incentive for conservation.

Within the five study regions, only the northeastern Colorado project offers assurance that supplies will be available to meet changing water needs or to satisfy high-value uses during periods of shortages. In the other areas, the institutions and incentives to encourage transfers among areas or alternative uses are lacking. The custom of not attaching a price to water provides no incentive for neighboring jurisdictions to cooperate in meeting the projected needs of Virginia Beach. Likewise, attractive opportunities for Kern County to buy water from its neighbors to the north and for the Metropolitan Water District to purchase water from the Imperial Irrigation District have yet to materialize because of institutional barriers to such trades. In the Columbia River basin, the favored treatment given offstream water uses threatens to divert water from its already established, high-value uses in hydropower production to a very marginal use in irrigation.

The authors of these water-problem studies offer recommendations for keeping supplies in balance with demand over the long term and for responding effectively to changes in water-use values. With institutional factors at the root of the problems, it is not surprising that institutional changes offer the most effective — and in some cases perhaps the only — solutions. Virginia Beach, with its proximity to relatively abundant, unused water supplies, is the only one of the study areas that can realistically expect to resolve its problems largely through traditional, supply-side solutions while eschewing institutional reforms. But, as the town's fears of future shortages demonstrate, even this solution is not certain. Virginia Beach has been deterred from attempting to develop the most cost-effective additional supplies because, with existing institutional arrangements, this would leave them dependent on the uncertain cooperation of their neighbors. Moreover, the lack of clear property rights to the water Virginia Beach hopes to import adds to their concern.

In the western study areas, there are essentially no unused water supplies available to meet growing needs. Not only have the low-cost means of augmenting supplies been exhausted, but the higher-cost alternatives require diverting water from other uses. Institutional changes that encourage demand-side management and reallocation of water from low-value to higher-value uses is essential in the West if the more serious concerns about water shortages are to be averted.

Uncertainties surrounding water rights and the future availability of water for certain uses are common problems, and there is general agreement among the authors of the case studies that the courts and special-interest legislation are not the best means of resolving them. There is not complete agreement, however, as to how they should be resolved. The authors of the western case studies favor greater reliance on markets and price incentives for allocating scarce supplies, whereas the authors of the Virginia Beach study favor greater centralization to coordinate the development and allocation of water resources among different jurisdictions. The differences in approach may reflect the very different circumstances and experiences of southeastern Virginia and the western sites rather than philosophical differences among the authors as to the proper role of markets and government in resource allocation. Shabman and Cox claim that water is not scarce enough in southeastern Virginia for water marketing to operate effectively.

The case studies agree on the need for greater flexibility in transferring water and on the advantages of unencumbered property rights in water for achieving this flexibility. The studies do not address the question of what will lead to the desired institutional reforms. Do problems have to get much worse before sufficient pressure arises to force the desired change? The state and federal responses to the 1976/77 drought in California indicate that extreme stress does indeed stimulate reform. Revisions in both state and federal laws and policies were made to facilitate transfers of water to the hardest hit users and to counter the worst effects of the drought. Unfortunately, many of the reforms facilitating transfers did not outlive the drought.

The case studies in this volume offer some hope that constructive reform may not have to await a crisis. California recently passed a law stipulating that rights cannot be lost when water is sold, and state officials are actively exploring ways to encourage transfers from low-value agricultural uses in the Imperial Valley to the Metropolitan Water District of Southern California. (Of course, interest in such transfers emerged only in the near panic that followed voter rejection of the legislation authorizing the Peripheral Canal.) Idaho already has a system of water banking, and the state was forced to take a broader look at water rights when its supreme court overturned the presumed right of irrigators to preempt hydropower rights to water. And finally, the study of the Colorado-Big Thompson Project demonstrates that federal water projects can be compatible with water markets.

Obstacles to the emergence of effective water markets vary widely among the study areas. Undoubtedly, the most intractable obstacles are those protecting powerful vested interests. In California, however, some of the policies restricting the opportunities and incentives for water transfers benefit no apparent constituency. Laws and policies preventing California's water districts from making a profit and limiting water uses of federal irrigation projects provide no clear benefits to any group. Their removal, on the other hand, would benefit both current users of the water who would gain flexibility in how they can dispose of their water and those who would like to purchase the water for higher-value uses. In such cases, educating the prospective beneficiaries should increase the prospects for such changes. Protecting third-party interests without unduly restricting water transfers is likely to be the most difficult challenge in establishing efficient markets in California. The case studies suggest it may be easier to handle third-party effects when the transactions involve water districts rather than private water rights.

In the Pacific Northwest, the problems of creating unambiguous, unattenuated property rights as a first step to effective water markets are extremely complicated. Among issues that must first be resolved are federal claims to Columbia River water for an irrigation project of questionable economic merit, treaty provisions granting fishing rights to Indians, potential interstate conflicts over a river that has never been the subject of an interstate compact, and the relative rights of offstream and instream users. These issues pit powerful interests against each other in conflicts that cannot be resolved without some groups losing rights, however tenuous the claims may be, to the region's waters. The conflicts are being resolved very slowly through the courts or by lobbying for special-interest legislation. As the value of water rises, the costs of this system of resource management are likely to increase. Indeed, the situation is likely to deteriorate in the absence of a major federal initiative to resolve outstanding Indian claims, the status of the Columbia River Irrigation Project, and water rights for federally controlled hydropower facilities. These issues, as well as the increasing potential for interstate conflicts, give the federal government considerable leverage over the situation. If used wisely, this leverage could provide the key to devising and implementing changes. Once it is widely recognized that the overall interests of the region are not well served by existing arrangements, it may be possible to gain the cooperation of the state governments in seeking change. The carrot of federal project money, along with the stick of threatening to challenge state-issued water rights with federal and Indian reserved water rights, could provide powerful incentives for all parties to cooperate in reforming restrictive, inefficient arrangements.

The Colorado-Big Thompson Project provides a precedent for a more-flexible approach to use of water delivered through Reclamation projects. It also offers lessons that might aid in designing water institutions elsewhere. Several factors contribute to the unusually effective markets in C-BT water. First, the markets could not have emerged if C-BT water were subjected to the constraints normally placed on the use of water from Bureau projects. In view of the growing scarcity of western water, it is unlikely that the purposes allegedly served by tying water to particular uses, locations, or farm sizes justify the costs imposed when water is locked into inefficient, low-value uses. The flexibility permitted in the use of C-BT water might serve as a precedent for all federal projects.

Two other aspects of the project—proportional rights and ownership of return flows—are more controversial as to their relative merits under other circumstances. Both play important roles in the operation of markets for C-BT water, but they are not without drawbacks. The use of proportional shares makes it possible to have homogeneous water rights even though the supply varies. This feature facilitates the transfer of rights within a market. But since users avert risk of low flows by purchasing additional shares under a proportional rights system, the authors of the C-BT case study argue that more water ends up in the short-term rental market and, therefore, in lower-value uses than would occur under a system of priority rights. However, if more C-BT water does indeed get diverted to lower-value uses than is optimal, this may owe more to the fact that owners of the water rights are discouraged from charging the market-clearing price on their short-term rentals. At any rate, the advantages of having homogeneous rights for operating a market are considerable and might well offset any alleged disadvantages in the rental market. Of course, this issue would be moot for the C-BT Project if the annual deliveries from the western slope of the mountains had not been subject to variable quotas.

Perhaps the most interesting feature of the C-BT Project is the ability to ignore third-party effects by retaining ownership over return flows. An efficient exchange must allow for third-party effects, but the process often adds so much to transaction costs that water exchanges become sporadic and uncertain at best. Developing institutions for efficiently handling such effects is a major priority for improving water use efficiency. An important message of the C-BT study is that there may be circumstances under which the benefits to be derived from unimpeded water exchanges justify ignoring their third-party effects by retaining or buying up the rights to return flows.

Two other lessons suggested by the C-BT case study are the advantages of having even a small margin of water that is easily transferable and the need to provide compensation to the basin of origin when planning for water imports. Even though the C-BT Project provides only 17 percent of the region's total water supply, the ease with which it can be transferred among uses enables it to play a disproportionately important role in meeting changing water needs. Areas such as Virginia Beach and southern California that are encountering opposition to their water-import plans from the basins of origin should note that even in the 1930s northeastern Colorado had to allow for the future interests of those within the Upper Colorado River basin before a transmountain water import plan was approved.

Each of the case studies offers insights into certain types of water problems that occur in many other areas. But it is hoped that the volume as a whole does more than offer solutions to specific types of problems. The volume is intended to improve understanding of the nature of emerging water problems and the implications for alternative approaches to solving them. This understanding is requisite to developing and adopting policies that will make water use more responsive to the changing demands society is placing on its scarce water supplies.

References

Frederick, Kenneth D., with James C. Hanson, 1982. *Water for Western Agriculture* (Washington, D.C., Resources for the Future).

Rogers, Peter. 1983. "The Future of Water," *The Atlantic Monthly* vol. 252, no. 1, pp. 80–92.

Solley, Wayne B., Edith B. Chase, and William B. Mann IV. 1983. *Estimated Use of Water in the United States in 1980* (Alexandria, Virginia, U.S. Geological Survey Circular 1001).

U.S. Water Resources Council. 1978a. *The Nation's Water Resources 1975–2000*, vol. 1 (Washington, D.C., GPO).

U.S. Water Resources Council. 1978b. *The Nation's Water Resources 1975–2000*, vol. 3, app. II (Washington, D.C., GPO).

2
Competition Between Irrigation and Hydropower in the Pacific Northwest

*Walter R. Butcher and Philip R. Wandschneider, with Norman K. Whittlesey**

The Pacific Northwest is blessed with abundant, high-quality water resources. The Columbia River system, which drains most of the region, has an average flow several times as large as the water demands for irrigation and other consumptive uses. Until recently, flows remaining in the river appeared to be more than adequate for navigation, anadromous fish passage, and hydropower generation. (Dam construction had, however, seriously eroded fish production.) Then electricity demands began to exceed the region's capacity for generating hydropower and it began to appear that river flows were not adequate to sustain fish migration. Conflict is now emerging between competing water uses, especially between new diversions and existing instream uses along the main stem of the Columbia and Snake rivers. Irrigators in particular find that they now must contend with demands for instream use by hydropower and anadromous fish production.

Competition between diversion and instream uses was not anticipated when the region's water laws, authorities, and initial distributions of water rights were established. Thus these institutions are not entirely prepared to resolve the emerging issues in the competition. As a result, there is increasing stress and pressure for changes within the region's water institutions.

In this chapter the economics of the hydropower-irrigation tradeoff

*Professor, associate professor, and professor of agricultural economics, respectively, Washington State University, Pullman, Washington.

is examined and the institutions that guide allocation of the water are reviewed. We will examine two cases that show the stresses and strains accompanying the competition between irrigation and hydropower for use of water. The first is a direct confrontation in southern Idaho, where the Idaho Power Company has sued hundreds of irrigators and the State Department of Water Resources, charging a "taking" of water that rightfully belongs in the Snake River where it can be used to generate electricity. In the second case the confrontation has not been so direct; however, there is growing debate about how a proposed 500,000-acre expansion of the U.S. Bureau of Reclamation's Columbia Basin Irrigation Project will affect the cost of electricity.

Water Supply and Use in the Columbia River Basin

Water Supply

The Columbia River originates in the Canadian Rockies, flows south into the United States, through eastern Washington and then west along the border between the states of Washington and Oregon. The Columbia is joined by several tributaries, the largest of which is the Snake River. The Snake originates in the Jackson Hole area of western Wyoming, flows westward across southern Idaho, north along the Idaho-Oregon border, and finally west again to join the Columbia in southeast Washington. (The major rivers are shown on a map of the Pacific Northwest in figure 2-1.) At the point where the two rivers join, the mean annual (average) flow in the Snake River is 46,000 cubic feet per second (cfs) and the mean flow in the Columbia is 114,000 cfs (table 2-1). Other tributaries that flow into the lower Columbia swell its average flow by the time it reaches the mouth to 240,000 cfs. Table 2-1 shows average flows at various locations throughout the river system and the variability of flows at Bonneville Dam, on the lower Columbia.

The Columbia River has a distinct seasonal pattern of flows. Late summer and midwinter flows may be as little as one-fifth of the mean. When the peak flow occurs during spring snow melt, water is stored in upstream reservoirs for release during low-flow periods. However, the upstream storage in the Columbia River system is small relative to annual flows and, therefore, can make only a small contribution toward evening the flow from one year to the next.

Water Quality

Water quality is generally quite good in the Pacific Northwest (Petke, 1980). On the major rivers, large volume and rapid flow provide ample

Figure 2-1. Columbia-Snake drainage system

capacity to assimilate waste, but problems sometimes arise where large municipal or industrial discharges enter a smaller stream during low-flow seasons. Large withdrawals can contribute to problems of water quality by depleting flows, but the Northwest does not have a serious problem with salt concentration in return flows from irrigation.

The most pervasive water-quality problem in the Columbia River system is that of high water temperatures caused by reservoirs that impede water flow, together with water uses that deplete flows or raise temperatures of returned waters. Higher temperatures reduce habitat quality and increase mortality for the salmonoid fisheries. Hydropower and irrigation contribute to the temperature problem; however, with the hydropower dams already in place, diversion of water to irrigation

TABLE 2-1. Water Supply and Streamflow in the Columbia and Snake Rivers

A. Mean flow at selected Columbia and Snake River sites

Stream	Location	Drainage area (sq. mi.)	Mean annual flow[a] (million acre-feet per year)	Mean annual flow[a] (thousand cubic feet per second)
Snake	King Hill (Upper Snake)	35,800	6.2	9
Snake	Ice Harbor Dam (Lower Snake)	108,500	33.3	46
Columbia	Priest Rapids	96,000	82.6	114
Columbia	Bonneville	240,000	128.4	177
Columbia	mouth	259,000	173.5	240

B. Variability in Columbia River flows at Bonneville

	Annual flows[a] (million acre-feet)	Momentary flows[b] (thousand cubic feet per second)
Maximum	179.1	1,240
Mean	128.4	177
Minimum	95.1	35

Source: Pacific Northwest River Basins Commission, *Water Today and Tomorrow* (Vancouver, Wash., 1979) vol. II, p. 3-29.
[a] Values based on the 1929–1958 base period adjusted to regulated conditions as of 1970.
[b] Observed values for the period of record, 1895–1975.

would have only a small effect on water quality. Therefore, issues of water quality are excluded from our analysis.

The principal uses of the Columbia River, for irrigation, hydropower, navigation, and anadromous fish passage, are summarized in table 2-2.

Irrigation

Irrigation consumes about 20 million acre-feet (maf) of water per year in the Pacific Northwest, which is more than 95 percent of total consumptive use. Almost 9 million acres of cropland and pasture are irrigated in Washington, Oregon, Idaho, and western Montana. The irrigation takes place mostly in the upper Snake, middle Columbia, and along tributaries such as the Clark Fork in Montana, Yakima in Washington, and the Willamette in western Oregon. Most of the irrigation water is supplied by diversion from rivers, but nearly a million acres are irrigated from deep wells. The area of heaviest groundwater pumping is in the Snake River plain in southern Idaho.

Irrigation plays an important role in the region's economy. Fruits, vegetables, and high-value field crops such as potatoes and sugar beets are grown almost exclusively on irrigated lands. Many of the irrigated crops are processed within the region, providing employment and income in the associated agribusiness as well as in agricultural production.

Hydropower

The Pacific Northwest benefits greatly from the large supply of cheap hydropower produced mostly at dams on the Columbia River and its major tributaries. Dams currently in place in the Pacific Northwest can generate 110 billion kilowatt hours (kWh) of electricity per year during critical, low-flow conditions in the river, and about one-third more with median flows.

The average cost of the hydropower at the dam is only about $0.003 per kWh. As long as the region relied almost solely on hydropower, retail rates, including distribution costs, averaged less than $0.015 per kWh, about half the national average. Customers responded by using about twice as much electricity per person as the national average, and other electricity-intensive industries gravitated to the region.

During the 1970s, the Pacific Northwest entered a transition from hydropower to thermal power. The first large thermal plant came on-

TABLE 2-2. Major Uses of Water in the Pacific Northwest Region, 1980[a]

River use	Units served	Water used
Domestic water supply	8 million persons	0.7 maf consumed
Electric energy	59 Columbia System dams	Streamflow at each dam
Firm capability[b]	110 billion kWh	(88 maf at the Dalles Dam)
Secondary capability[c]	35 billion kWh	(40 maf additional at the Dalles Dam)
Irrigation	8.8 million acres	20 maf consumed
Barge transportation	8 million tons	Minimum flows at locks (16 thousand cfs, Lower Snake River)
Anadromous fish passage[d]	800,000 salmon/steelhead Columbia-Snake System	Spring reservoir flow through (110,000 cfs, Middle Columbia) (85,000 cfs, Lower Snake)

Source: Pacific Northwest River Basins Commission, 1979. *Water Today and Tomorrow* (Vancouver, Wash., 1979) vol. II, chapter 3.

[a] Idaho, Oregon, Washington, and western Montana. Except as noted, includes coastal streams in addition to the Columbia River.

[b] Pacific Northwest generation capability under critical (low-flow) conditions.

[c] Additional generation possible with median flows.

[d] Based on 1970–1975 runs. Indicated flows are daily averages, recommended along with other changes to restore runs to pre-1960 levels.

line in 1971. The last major power dam in the region was added in 1975 (Lee and coauthors, 1980). By 1980, hydropower supplied only 75 percent of the region's firm energy; however, it still supplied all of the peaking capacity.

The high capital and operating costs of thermal power (estimated to be more than ten times the cost of hydropower) caused wholesale and retail electricity rates to increase sharply (Northwest Power Planning Council, 1983). By 1983, the wholesale rate charged by the Bonneville Power Administration (BPA) for federal system power was $0.025 per kWh, pushed up from only $0.005 per kWh in 1978.

Rate increases were followed by declines in sales as customers responded to the higher costs. The projected demand for power has not materialized and the region now has a firm power surplus of 1,500 MW. Six of nine nuclear power projects that were being pursued in 1979 have been canceled and construction on two more plants has been suspended. The Washington Public Power Supply System defaulted on $2.25 billion in bonds issued to finance construction at two of the canceled plants. Other utilities have also encountered financial difficulties. Some of the region's utilities are now trying to negotiate long-term sales of plants or power outside the region.

The region's experiences with thermal power have heightened competition for the cheap hydropower from the Columbia and Snake rivers. The region's public and private utilities, the aluminum companies, and the California utilities that buy surplus power from the Northwest all are seeking to secure rights to the cheap federal hydropower and to shift the obligation for the cost of thermal power to someone else.

Fisheries

Before development on the Columbia River system began, millions of salmon and steelhead trout traveled each year from the Pacific up the Columbia River to spawn in its tributaries. Depletion of flows and blockage by upstream dams have cut off many of the traditional spawning areas. The major hydroelectric dams on the main stem of the Columbia and Snake rivers have further depleted the salmon runs through damage to juvenile fish as they move out to the ocean and through impediments to returning adult spawners. These and other changes such as overharvesting have reduced runs on the Columbia to approximately 500,000 fish per year above Bonneville Dam. A new program to restore a portion of the fish runs grants fishery agencies rights to a large instream flow during the spring downstream migration of the young fish.

Navigation

Oceangoing vessels travel 100 miles up the Columbia to Portland, Oregon. Barge traffic moves through a series of locks and reservoir pools up the Columbia to Pasco, Washington, and from there up the lower Snake to Lewiston, Idaho. The barges carry mostly grain downriver to Portland and petroleum and fertilizer back to upriver terminals.

Since navigation takes place through pools and locks, minimum channel depth can be achieved at much less than normal flow. A flow of 16,000 cfs at Ice Harbor, the last dam on the lower Snake, is sufficient for navigation purposes.

Water and Energy Tradeoff

During most of the Pacific Northwest's history, the major uses of water from the Columbia and Snake rivers were assumed to be largely compatible. Energy from hydropower was abundant and cheap. Water withdrawals for several million acres of irrigation caused no alarming depletion of flows in the rivers. There appeared to be adequate streamflow for the passage of migratory fish through the fish ladders at downstream dams (although other dams had cut off traditional spawning areas on the upper rivers).

From 1960 to 1980, the increase in irrigation and the building of hydropower capacity both accelerated. By 1980 the hydropower system could generate twice as much electricity from a given volume of water. In the same 20 years, the area irrigated and the volume of water withdrawn for irrigation increased by about 30 percent, and additional irrigation and attendant water depletion were anticipated. To benefit fish, a strong case was made for shifting streamflows to the spring, despite the adverse effect on hydropower production. Since there were no further opportunites to build new dams, the level of future hydropower production became entirely dependent on the amount of streamflow. The decision to increase irrigation thus came to imply not only the usual costs of delivering water and farming the land but also the costs of hydropower production lost through the depletion of streamflows.

Hydropower Production

Hydropower production is determined by developed head (that is, the height of a retained body of water) above generating turbines and the volume of water flowing through the turbines. Figure 2-2 shows the

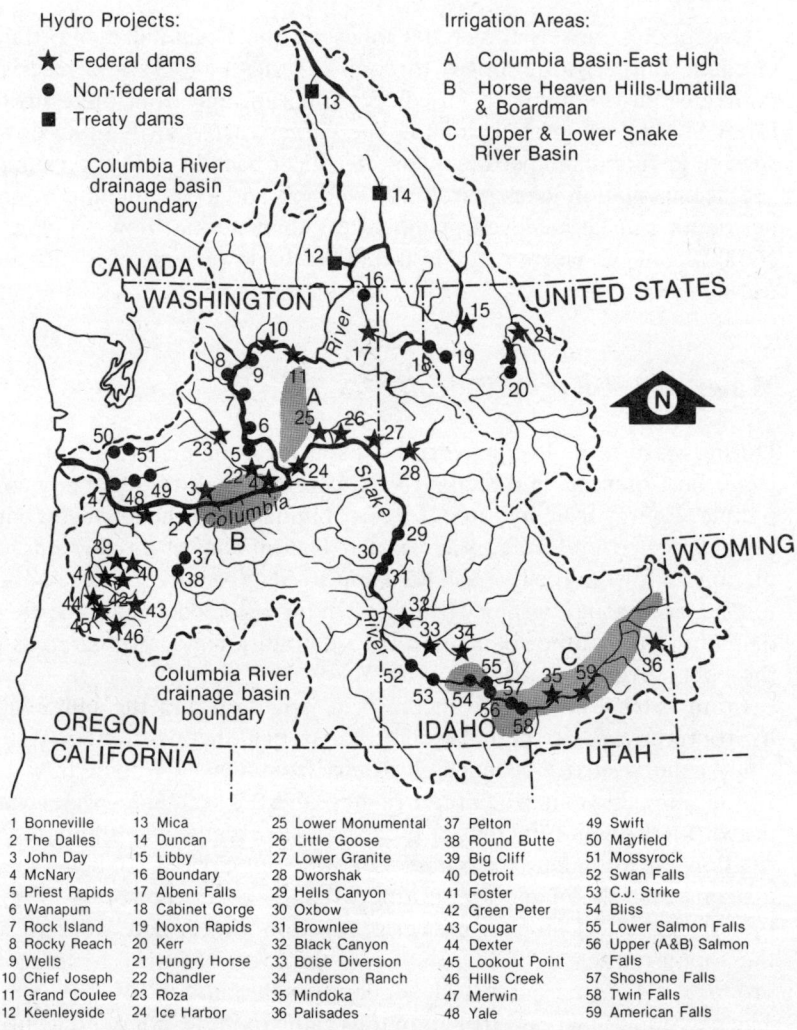

Figure 2-2. Columbia-Snake River basin: hydropower projects and primary future irrigation areas

location of major hydropower projects in the Columbia and Snake river basins. Developed head and generating capability for power structures on the Columbia and Snake rivers are listed in table 2-3. Cumulative head is the sum for all structures downstream from a particular dam. Cumulative generating capability is based on 0.87 kWh generated for 1

TABLE 2-3. Developed Head and Generating Capability at Hydroelectric Power Structures on the Snake and Columbia Rivers, 1960 and 1980

River and structure	Gross head per dam (feet)	Cumulative head to sea (feet) 1960	Cumulative head to sea (feet) 1980	Cumulative generating capability[a] (kWh/a-ft) 1960	Cumulative generating capability[a] (kWh/a-ft) 1980
Columbia River, Lower (Washington-Oregon)					
Bonneville	59	59	59	51	51
The Dalles	83	142	142	124	124
John Day	100	—	242	—	211
McNary	74	216	316	188	275
Columbia River, Middle (Washington)					
Priest Rapids	77	—	393	—	342
Wanapum	77	—	470	—	409
Rock Island	34	250	504	218	438
Rocky Beach	87	—	591	—	514
Wells	67	—	658	—	572
Chief Joseph	167	417	825	363	718
Grand Coulee	342	759	1,167	660	1,015
Snake River, Lower (Washington)					
Ice Harbor	98	216	414	188	360
Lower Monumental	100	—	514	—	447
Little Goose	98	—	612	—	532
Lower Granite	98	—	710	—	618
Snake River, Middle (Idaho-Oregon)					
Hells Canyon	210	—	920	—	800
Oxbow	120	—	1,040	—	905
Brownlee	272	488	1,312	425	1,141
Snake River, Upper (Idaho)					
Swan Falls	24	512	1,336	445	1,162
C. J. Strike	88	600	1,424	522	1,239
Bliss	70	670	1,494	583	1,300
Lower Salmon Falls	59	729	1,553	634	1,351
Upper Salmon Falls "A"	46	775	1,599	674	1,391
Upper Salmon Falls "B"	37	—	1,636	—	1,423
Shoshone Falls	214	989	1,850	860	1,610
Twin Falls	147	1,136	1,997	988	1,737
Minidoka	48	1,180	2,045	1,027	1,779
American Falls	49	—	2,094	—	1,821

Source: Norman Whittlesey, Joanne R. Buteau, Walter R. Butcher, and David Walker, *Energy Tradeoffs and Economic Feasibility of Irrigation Development in the Pacific Northwest*, Washington State University, College of Agriculture Bulletin 0896, 1981.

[a] Generating capability may be slightly reduced during seasons when reduced water levels behind storage dams such as Grand Coulee may decrease generating head.

acre-foot of water falling through 1 foot of head. The cumulative generating head ranges from 59 feet, above Bonneville Dam on the lower Columbia River, to as much as 2,094 feet above the upper Snake River dams in eastern Idaho. The electricity that can be generated by 1 acre-foot of water flowing downstream from each of the locations ranges from 51 kWh per acre-foot at Bonneville to 1,821 kWh per acre-foot at American Falls.

The head and the generating capability on the Columbia and Snake rivers almost doubled between 1960 and 1980. Further development on the rivers is unlikely, however, as the few remaining undeveloped reaches of these rivers have been withdrawn from development for environmental or other reasons. Thus, future hydropower production will depend on the instream flows remaining after water depletions upstream.

Opportunity Cost of Water Depletion

Depletions from streamflows cause an opportunity cost (that is, value forgone) equal to the value of energy that could have been produced had the water been left to flow downstream through the hydropower generators. Depletions occur when water is diverted and either consumed or diverted around hydropower dams before flows return to the river.

During most of the Pacific Northwest's history the opportunity cost of streamflow depletions has been small. As shown in table 2-3, only about half the present hydropower generation capability was installed by 1960.

Moreover, the system lacked reservoir and turbine capacity for exploiting the high spring and summer flows. During the spring runoff, water was usually spilled over the dams without generating any power. Thus, depletions during spring and early summer generally did not reduce annual electricity output except in an unusually dry year.

From 1960 to 1980, completion of several major dams on the Columbia and Snake rivers and installation of additional capacity to make use of high flows greatly increased the amount of electricity that could be generated with a given amount of water. Also, the value of electricity increased from about $0.0025 per kWh in 1960 to $0.035 per kWh in 1980. Net hydropower loss and opportunity cost per acre-foot of water, shown in table 2-4, increased severalfold. In southeast Idaho the cost of replacing energy lost through depletion rose from $1.70 per acre-foot in 1960 to $64 per acre-foot per year in 1980. Other areas experienced similar increases in the economic cost of replacing the hydropower loss through depletion of river flows (Whittlesey and coauthors, 1981).

TABLE 2-4. Opportunity Cost of Diverting Water from Hydropower Production, Selected Pacific Northwest Areas, 1960 and 1980

Diversion area	Cumulative head (feet)	Energy loss[a] (kWh/a-ft)	Opportunity cost[b] ($/a-ft)
A. 1960 hydropower development			
SE Idaho (American Falls)	1,184	680	1.70
SW Idaho (Swan Falls)	512	256	0.64
Lower Columbia (The Dalles)	142	66	0.17
Columbia Basin (Grand Coulee)	759	272	0.68
B. 1980 hydropower development			
SE Idaho (American Falls)	2,094	1,822	64.00
SW Idaho (Swan Falls)	1,336	1,162	41.00
Lower Columbia (The Dalles)	242	211	7.00
Columbia Basin (Grand Coulee)	1,167	1,015	36.00

Source: Norman K. Whittlesey, Joanne R. Buteau, Walter R. Butcher, and David Walker, *Energy Tradeoffs and Economic Feasibility of Irrigation Development in the Pacific Northwest*, Washington State University, College of Agriculture Bulletin 0896, 1981, tables 8, 9, and 10.

[a] Based on 0.87 kWh/a-ft/ft of head. For 1960, it is assumed that 70 percent of diversions were lost to power production and 90 percent of return flow provides generation. For 1980, virtually all net depletions cause power loss.

[b] For 1960, $0.0025/kWh and for 1980, $0.035/kWh estimated "avoided" cost of the least-cost alternative (Office of Applied Energy Studies, 1982).

Irrigation Impact on Hydropower

As the power situation has changed, concern has risen over the impacts of irrigation on instream flows and hydropower production (Schuy, 1975; Department of Agricultural Economics, 1976). Irrigation interacts with hydropower production in three ways:

(1) Irrigation changes the seasonal pattern of electricity demand, streamflows, and hydropower supply.
(2) Irrigation consumes large amounts of electricity for pumping.
(3) Irrigation depletes streamflows and thus reduces downstream hydropower generation.

Seasonal Patterns. Part of the water withdrawal and power use for irrigation takes place in the spring and early summer when streamflows are high and power demands, which are heavily affected by winter heating demands, are low. Later, in the fall and winter, a portion of the water percolates through the groundwater and returns to the rivers at a time when streamflows are lower and other power demands are higher.

At one time, the seasonality of irrigation and the hydropower system fit together so well that the use of water for irrigation and for power during the early part of the irrigation season cost the power system nothing—no hydropower generation lost and no power diverted from other markets—during years when spring runoff was normal or above. Now, however, increased capacity for storing spring runoff, additional turbines in the dams, and new uses for summer power (such as irrigation pumping, displacing thermal plants, and selling to California) make it possible to use all of the annual streamflow for power generation except in an unusually wet year.

Irrigation Energy Requirements. Before 1960, most irrigation water in the Pacific Northwest was conveyed and applied to the land by gravity flow. In contrast, most of the irrigation developed during the past 25 years and almost all potential developments use electric pumps to lift water and to pressurize sprinkler systems. The 2.2 million acres under active consideration for development by the year 2000 would require an estimated 5.4 billion kWh of electricity per year for pumping, more than half as much as the 1980 electricity load for irrigating 8 million acres (Whittlesey and coauthors, 1981).

Irrigation demands for electric energy are of concern to planners in the region because of the magnitude of the demands and because of the substantial difference between the price paid by irrigators and the incremental cost that will be incurred by the power-production system. Private irrigators, under existing pricing policies, obtain power at an average cost of about $0.02 (20 mills) per kWh, which is about twice the rate of five years ago but only half of the long-run incremental (marginal) cost of power from new generating sources (Northwest Power Planning Council, 1983).[1] There is a search for ways to reduce the price of power for irrigators (Bonneville Power Administration, 1984); however, it is recognized that a price less than marginal cost is likely to encourage irrigators to use electricity inefficiently (College of Agriculture Research Center, 1981).

Hydropower Loss. The loss of hydroelectric energy when water is withdrawn and consumed for irrigation depends on the quantity of water

---

[1] Power for pumping at U.S. Bureau of Reclamation projects is priced at only $0.001 per kWh, far below even the average cost.

diverted and consumed and the amount of developed head that the water would fall through if left in the stream.

For potential new irrigation areas, the net depletion of water is expected to range from 1.5 to 3 acre-feet per acre per year. The hydropower loss per acre-foot reduction in flow was shown in table 2-4 to range from 211 to 1,822 kWh per acre-foot. The net hydropower loss per acre irrigated ranges from about 500 kWh per acre along the lower Columbia up to nearly 4,000 kWh per acre per year in the highest irrigation areas in eastern Idaho (Whittlesey and coauthors, 1981). Over the entire 2.2 million acres of land considered for irrigation development within the next 20 years, the energy loss averages 2,400 kWh per acre per year. The total hydropower loss to the region if all of this land were developed would be 5.3 billion kWh per year. By comparison, the output from a 1,200-MW thermal plant averages about 6.8 billion kWh per year. The 8.8 million acres under irrigation in the region result in a reduction of 17 billion kWh per year in Columbia-Snake hydropower generation (Houston, 1985).

The economic value of the hydropower loss depends on the cost of replacing the power. At $0.035 kWh, which is the cost in 1985 of the least expensive conservation options in the region, the hydropower loss would average $85 per acre per year. Most of the generation that is expected to be added in the next 20 years is expected to cost about $0.055 kWh, which would make the hydropower loss equal to $134 per acre per year.

Value of Water in New Versus Existing Irrigation

A large opportunity cost for depletion of irrigation water does not mean that irrigation should cease. Irrigation that can yield more value than hydropower should have use of the water. Continued allocation of water to existing irrigation is particularly likely to yield high value in return on irrigation investments that have already been committed. Capitalized values of about $2,000 per acre indicate a value of water in existing irrigation that is more than the marginal hydropower value of water from even the highest elevations in the river system.[2] The net

---

[2]Lands used in established orchards and other high-value crops yield high returns for water. But there are also much larger irrigated acreages in the Pacific Northwest, where reallocation would be appropriate because the marginal value of water is low for the following reasons: (1) low-productivity land used only for low-valued irrigated pasture or small grain production, (2) high non-irrigated value of land with enough rainfall to produce a good crop of grain, (3) high operating costs on land that requires a large amount of pumping power, or (4) opportunities for marginal reductions in water consumption with only a small impact on net income.

regional economic value of irrigation would decrease if water from these lands were returned to the river for use in hydropower production.

For potential new irrigation the situation is quite different. Investments have not yet been made in water diversion and irrigation systems. The marginal value of water in new irrigation thus is reduced by these costs that are incurred if, and only if, the water is allocated to irrigation. In a 1981 study of irrigation feasibility in the Pacific Northwest, Whittlesey and coauthors (1981) estimated development costs ranging from about $1,000 to $3,500 per acre. For most of the potential development areas, the present value of net returns to land and water, adjusted for income tax benefits that could be realized by a private investor, was less than the investment in undeveloped land and the irrigation system.

There are other indications that the net value of water, at point of withdrawal, is approximately zero for new irrigation development. On theoretical grounds, irrigation developers can be expected to develop new projects as soon as technological and economic conditions create an expectation of net returns above costs of development and operation. In other words, developers will expand irrigation whenever the marginal value of the water at the point of diversion promises to exceed the marginal cost to the irrigator, which is zero under current laws. Prices of developed irrigated lands are about equal to the cost of developing new lands, indicating that there is no residual value of water left to be capitalized into the value of the land.

Irrigation-Hydropower Tradeoff: Summary

In summary, a conflict is emerging between irrigation and hydropower because changing technological and economic conditions have greatly increased the value of water for energy production. The power that can be generated from water in the upper reaches of the Columbia and Snake rivers is worth from $35 to $64 per acre-foot per year (see table 2-4). Electric utilities and consumers are growing increasingly protective of current flows against withdrawals that would force them to incur replacement power costs of $35 per acre-foot of depletion in river flows. Irrigation developers, on the other hand, face an economic situation in which they can seldom afford to develop land for irrigation even when the water is available to them without charge (at the point of diversion). Forcing developers to take the hydropower value of water into account would preclude almost all new irrigation development in the region. As this conflict of interest emerges, the power interests are seeking means to assert and defend their rights to instream flows of water and the irrigation developers are seeking to preserve their traditional right of free access to additional water to serve new irrigated lands.

Actors and Institutions

Although economics provides an objective, or standard, for allocation of water, the actual outcome of competition between hydropower and irrigation uses of water depends on water institutions and significant actors. In this section we review state, federal, and interstate institutions that guide the allocation of water between hydropower and irrigation. The institutional background is important for the Swan Falls and Columbia basin case studies that follow.

State Water Institutions

In general, individual (consumptive) water users obtain access to water through water rights granted under state water laws. In the Pacific Northwest, state water law is constructed around the doctrine of prior appropriation.[3] Real property rights are granted to the use of the water. Historically, water rights were "self-created" by the simple act of claiming water and putting it to use. Most state water laws now require that the prospective user apply for a water use permit from a state water agency and then proceed to use the water beneficially. Riparian land is not necessary nor does it guarantee access to the water. Rights may be bought and sold, though transfer is restricted to protect third parties.

One feature of the appropriation doctrine is the chronological principle governing priority of access: first in time is first in right. Prior appropriators (literally, first takers) are entitled to use their full allotment of water before later (junior) claimants use any. Shortages are not shared pro rata but are borne entirely by junior rights holders. Therefore, the date attached to a right can become critical in conflicts, as in the cases involving Swan Falls and the Columbia basin. Generally, the water right bears the date of application to the water agency, not the date of first use of water.

Another feature of the appropriation doctrine is the principle of beneficial use: water rights are limited to amounts "reasonably required" to meet a "beneficial use." Applications for excessive amounts of water will be denied initial water permits, and water not being used may be forfeited. The potential for loss through forfeiture is a disincentive to water conservation (since saved water may simply revert to the public domain) and to temporary or partial water transfers (since rental of water may be construed as evidence that the water is surplus to needs).

Beneficial water uses are defined by state constitutional and statutory law and case law. Early water codes (for example, the Water Code of

---

[3]See Hutchins (1971, 1974, 1977) for a comprehensive survey of western water law.

1917, Washington; and the Acts of 1891, 1899, and 1905, Oregon) were concerned with the diversion of waters. Perfection of water rights typically required the construction of facilities to divert or impound the water, followed by the beneficial use of the water. Clearly this terminology did not contemplate rights for instream natural-flow uses such as hydropower and fishery habitat. However, recent state laws have expressly declared some instream uses beneficial and have established procedures by which the state water agencies allocate water for these purposes, usually through minimum flows or reservations of water from appropriation (for example, Washington's 1971 Water Resources Act, see RCW 90.54.020).

Hydropower projects use flowing water but also generally involve some physical water-control facilities and often are associated with storage reservoirs. Thus, the status of hydropower water rights lies between that of irrigation (a clear diversionary use) and fish habitat (a natural-flow use). The natural flow is modified, but the water is not diverted and "consumed." In general, states do issue rights to flows at individual hydropower projects, but earlier rights were often explicitly subordinated to diversionary uses such as irrigation and municipal water supply.

Federal Water Laws and Actors

The state's authority over water is shared with the federal government.[4] By virtue of the Supremacy Clause of the Constitution, where state and federal laws conflict, federal laws must prevail. However, federal authority is limited to the powers enumerated in the Constitution (as interpreted by the Supreme Court). The federal government does not generally license water allocations to private parties and regularly includes "saving" provisions to protect state-granted rights. There is no federal water code analogous to state water codes. Instead, there are relatively independent bodies of federal law related to particular water agencies (for example, U.S. Bureau of Reclamation and the U.S. Army Corps of Engineers), water development projects (for example, the Columbia Basin Project), and water uses (for example, environment, power, navigation, irrigation, flood control).

The U.S. Bureau of Reclamation has major authority for federal irrigation development. Two features of the laws regulating the Bureau's activity are especially relevant to the irrigation-hydropower conflict. The first is that the Bureau's capital expenditures for irrigation projects are

---

[4]See Hutchins (1971, 1974, 1977) for water law. See Holmes (1972, 1979) for an overview of federal water programs and policies. See Blumm (1983), and Norwood (1981) for an overview of federal action in the Northwest, focused on hydropower.

subsidized by provision for repayment of federal capital expenditures at zero interest over a fifty-year period and by a "basin account,"[5] which uses revenue from sales of electric power to pay a portion (71 percent in 1979) of the capital costs of irrigation construction (Norwood, 1980).

The relevant feature of reclamation law is that the Bureau of Reclamation is directed to follow the normal procedures of state water law to obtain legal access to water.[6] For example, the Bureau holds water rights and applications dated to 1938 for the completed and planned portions of the Columbia Basin Project in the Big Bend area of the mid-Columbia. In general, the Bureau holds the water rights and delivers water to irrigation districts and individuals under terms of a repayment contract.

The other major federal water development agency in the Northwest is the U.S. Army Corps of Engineers. The Corps has a wider mandate than the Bureau of Reclamation and an orientation toward river management; Congress grants it general authority over navigation and flood control (Rivers and Harbors Act of 1899 and Flood Control Act of 1944). The Corps is also the major producer of hydropower in the Northwest. Based on the constitutional superiority of national law, the Corps does not apply for state water rights or other state permits for its projects. Instead, the Corps proceeds under its general mandate over navigation and on the basis of the separate congressional acts that authorize specific projects.

The Federal Energy Regulatory Commission (FERC) is the primary agent of federal authority over nonfederal hydropower projects. In effect, FERC licenses allocate water rights to private parties for hydropower production. The Supreme Court has ruled that federal authority to license nonfederal dams cannot be conditioned by state power (*First Iowa Hydro Electric Coop* v. *FPC*, 328 U.S. 151 [1946]). Conflicting state law must give way to a FERC license.[7] FERC licenses contain provisions regulating the operation of licensees. Such provisions may include minimum streamflows for purposes other than power production, such as fish habitat, recreation, or flood control.

In addition to these two water development agencies, the federal government charters a number of other agencies that have water-related responsibilities in the Northwest. The Bonneville Power Administration

---

[5]P.L. 89-561, 16 U.S.C. 835 (1964).

[6]Section 8 of the Reclamation Act of 1902, upheld by the *Supreme Court in California* v. *the United States*, 438 U.S. 645 (1978). However, where state law conflicts with a congressional directive, federal law prevails.

[7]See *State of Washington Department of Game* v. *FPC*, 207 F. 2d 391 (9th C.A. 1953), cert denied 347 USA 936 (1954), in which an FPC license to build a dam prevailed over state law to bar it.

(BPA) markets almost all federally produced (Corps of Engineers and Bureau of Reclamation) power in the Northwest. The role of BPA in marketing includes the scheduling of power from these sources to meet its loads. This scheduling function implies a large degree of control over the operations of hydropower generators and power-storage reservoirs. BPA is therefore a major actor in the allocation of streamflows over time.

The Pacific Northwest Power Planning Council (NPPC) is a unique federal-state agency. It was established in 1980 by Congress, but council members are appointed by the governors of Idaho, Montana, Oregon, and Washington. The council is funded through BPA, but its main responsibility is to determine plans and policies that are in the best interest of the region. The council's authority for fish and wildlife planning has thrust it into the issue of water allocation on the Columbia River. It has established a "water budget," which assigns a block of Columbia-Snake river water to fishery interests from the water otherwise available to be managed by the hydropower agencies.

Another important feature of federal water law is legislation authorizing and appropriating money for federally sponsored water projects. The authorization and appropriation acts specify the purposes of the projects and outline dimensions and constraints. While Congress authorizes the investment, it does not generally allocate water for the project, that being a state function.

The final class of federal water law comprises other, more general laws, which bear on water use and management. Among such laws relevant to Columbia River water allocation and the irrigation-hydropower tradeoff are (1) laws establishing land subsidies for irrigation (the Desert Land Act, 1877; the Carey Act, 1894), which have abetted the development of irrigation especially in Idaho (Chaney, 1977); (2) laws governing management of the anadromous fishery (for example, the Fish and Wildlife Coordination Act, 1934) and requiring federal agencies to consider fishery impacts in their decision-making; and (3) environmental laws that impose procedural requirements (the National Environmental Policy Act, 1969, requiring environmental impact statements) or impose substantive constraints (rivers withdrawn from development under the Wild and Scenic Rivers Act, 1968).

Interstate Water Allocation Institution

The Columbia-Snake river system passes through several states, so allocations of water in one state affect the use of water in other states. For example, consumption of water by irrigation in southeast Idaho reduces the instream flow for fish and hydropower production in the downstream states of Oregon and Washington. Interstate water allocation falls within neither state nor federal basic water law structure. Ultimately, interstate water law is subject to the Constitution, but the quasi-sovereign status of states implies that interstate allocation is not a simple subcategory of federal water law. There are three institutional formats in which interstate water allocation issues may be considered: interstate compact, federal legislation and agency action, and adjudication by the Supreme Court.

An interstate compact is somewhat analogous to a treaty. In general, interstate water compacts have been limited to allocation of water quantities to the participating states, and rarely possess regulatory power over specific water uses. There is no Columbia River compact, despite discussions over many decades.

Federal legislation and agency action are limited only by the restraint of Congress and the Constitution. Rarely does it expressly allocate water between states (the Boulder Canyon Act is a major exception). The interstate aspects of the federal laws just discussed illustrate the typical federal influence on interstate water allocation.

The Supreme Court has developed a body of interstate and federal case law for water allocation. The fundamental principle is "equitable apportionment." This principle follows neither riparian nor appropriation doctrine, but requires that each state sharing waters obtain a "fair share," determined on a case-by-case basis. Recently the court has addressed questions of water transfers across state lines. To date, the court's rulings have not been definitive (Utton, 1983). In some cases, the court rules under the Commerce Clause that states cannot prohibit the transfer of water (or any other commodity) across state lines. At other times, the court rules that a state's authority to allocate water allows it to prevent the transfer of water out of state. The court has made no significant rulings concerning interstate allocation of the Columbia River.

Actors and Institutions: Conclusions

Water laws and agencies in the Northwest are governed by two major institutional structures. On the one hand, state water law governs water allocation with an emphasis on uses that divert and consume water. On the other hand, federal legislation delegates water rights to federal agencies and licenses and regulates private and other public users with an emphasis on instream uses.

State and federal institutions each include some consideration of both consumptive and instream uses, and there is coordination between state and federal institutions; but there is no systematic integration of federal and state law. Thus, there is opportunity for conflict, especially between consumptive uses and instream uses. The first case study, Swan Falls Dam in Idaho, involves an irrigation-hydropower conflict that has arisen mostly within state jurisdiction. The second case, that of the Columbia Basin Project, involves more directly the tension between state and federal law.

The sharing of water authority between state and federal governments also leaves interstate water allocation to be determined simply by the outcome of individual state laws and federal laws. At this time, interstate water flows are subject to no integrated and purposeful management. Since no compact negotiations, lawsuits, or federal laws currently bear directly on the interstate issue, it is not the subject of a detailed case study.

The Swan Falls Case in Idaho: Intrastate Irrigation-Hydropower Conflict

The Swan Falls case[8] represents the first explicit confrontation between hydropower production and irrigation development in the Pacific Northwest. The substance of the case is a suit in Idaho state courts by the investor-owned Idaho Power Company to defend its state-granted water rights at Swan Falls Dam on the Snake River from encroachment by water diverters throughout most of southern Idaho. Water rights for Swan Falls bear an early priority date, but were widely believed to be subject to preemption by irrigation and other diversionary uses. Therefore, the state has permitted the apparently conflicting use by irrigators.

Irrigation and Hydropower in Idaho

It is surprising that the confrontation between irrigation and hydropower would first occur in Idaho. Both are highly regarded industries that have each benefited from the other's presence in the state. Irrigation is relatively more important for the state of Idaho than for the other Northwest states. The area of land under irrigation in Idaho is about equal to the combined area of irrigated lands in Oregon and Washington.[9] Furthermore, the Idaho economy has limited industrial activity, so irrigated agriculture holds a prominent position in the economic base

---

[8]See Hamilton and Lyman (1983) and *Idaho Power Co.* v. *State*, 104 Idaho 575, 661 P. 2d 741 (1983) for background, facts, and legal issues.

[9]In 1980 Idaho had 4.0 million acres, Oregon 2.3 million acres, and Washington 2.0 million acres of irrigated land (Pacific Northwest River Basins Commission, 1981).

for the state and many of the local communities. There is long-standing political support for expansion of irrigation. Idaho state government has taken an active role in helping irrigators gain access to land and water, and the State Department of Water Resources looks favorably on irrigation, as evidenced by its optimistic water plans and liberal permit-issuing policies.

The Idaho Power Company also plays a prominent role in the economy and politics of Idaho. Idaho Power has been the principal supplier of electricity in southern Idaho since the turn of the century. The main source of its power has been several hydroelectric dams on the Snake River and its tributaries. A large part of its power sales have been to irrigators and to customers who depend on irrigated agriculture.

The relationship between irrigation and power in Idaho is evident in an agreement reached when the Idaho Power Company and the federal government competed for rights to dam the middle Snake River in the Hell's Canyon area. Idaho Power prevailed, at least in part, because of strong backing from the governor and the state's congressional delegation. A factor in gaining that political backing was the company's agreement to subordinate hydropower water rights at the middle-Snake dams to future withdrawals for irrigation. The company could also point to its record of encouraging the development of irrigation and supplying cheap power to irrigators.

Despite this history of compatibility between irrigation development and hydropower, the basic forces that led to the eventual confrontation were building throughout the late 1960s and the 1970s. The Idaho Power Company was repeatedly blocked in its attempts to increase its power generation. First, the plan of a consortium of utilities to build a power dam on the middle Snake River (the High Mountain Sheep Project) was overturned by the U.S. Supreme Court in 1967, and then the site was permanently lost through inclusion of that reach in the Wild and Scenic Rivers System. After going outside the state to buy shares in a coal project (Jim Bridger), Idaho Power was again rebuffed when the proposed Pioneer coal plant, near Boise, was rejected by the Idaho Public Utility Commission. Then, in 1979, the company's plans to build the Guffy Project on the Snake River above Hell's Canyon were abandoned because of environmental concerns.

During the time that Idaho Power was experiencing these difficulties with expanding its capability to generate electricity, a boom was occurring in irrigation based on high-lift pumping. During the 1960s and the early 1970s, some 40,000 acres per year were developed for irrigation in southern Idaho. It appeared that vigorous development might continue.

The Carey Act lands in southern Idaho were reopened to irrigation development in 1973. By 1977, 600,000 acres had been filed on under the Carey Act and an additional 400,000 acres under the Desert Land Act (Chaney, 1977). Water permits had been issued and were outstanding for large amounts of new irrigation on lands that had not yet been developed. Furthermore, there was suspicion of a substantial number of unauthorized water withdrawals that had never been officially filed. Finally the State Water Plan (Idaho Water Resources Board, 1982), which was first issued in 1976, called for *minimum* development of irrigation by the year 2000—850,000 acres for full irrigation and an additional 255,000 acres of supplemental irrigation in the Snake River basin.

If the acreages called for in the State Water Plan were all developed, the flow of water in Snake River would be reduced by about 1.7 million acre-feet per year. Chaney (1977) estimated that this would decrease power generation by 824 million kWh per year. Thus, by the mid-1970s there were indications of further expansion of irrigation and the accompanying water depletions at a time when Idaho Power was experiencing increasing difficulty in finding new power resources. Whereas changes in the national economy (high interest rates, low product prices, and high energy costs) subsequently slowed the expansion of irrigation, they did not eliminate the emerging competition for water.

The Swan Falls Case

Concern over water diversions for irrigation first surfaced in the Idaho Public Utility Commission (PUC), which regulates the prices utilities may charge. Rate payers (customers) of the Idaho Power Company complained to the commission that irrigation diversions were causing electricity rates to be unnecessarily and unjustly increased. The power company was being forced to build new facilities and produce costly thermal power simply because the generation capability of existing facilities was being eroded by lost streamflow.

The Swan Falls Dam proved a convenient although somewhat surprising vehicle for translating the rate payers' complaint into specific action. The Swan Falls plant is a relatively minor hydroelectric producer. It can make use of only a fraction of the flows that occur during the high-flow season and generates on average only 80 million kWh per year. However, Swan Falls possesses two features that led to its pivotal role. First, it is strategically located downstream from the farming area in southern Idaho and just above the Snake River canyon, which contains three major Idaho Power Company hydroelectric facilities (see map, figure 2-2). Therefore, a prohibition against additional diversions upstream from tiny Swan Falls Dam would effectively protect flows for Idaho Power's huge middle-Snake dams[10] where subordination clauses preclude any direct defense of flow rights. Second, the water rights granted by the state of Idaho for 8,400 cubic feet per second of flow for power generation at the Swan Falls Dam hold priority dates from 1907 to 1930. Therefore, they predate most of the irrigation water rights.

The specific complaint to the Idaho Public Utility Commission was that Idaho Power had failed to defend its water right at Swan Falls against upstream diverters. The power company thus had allowed an illegal transfer of private property (that is, the water right) to the irrigators without having obtained either (1) permission from the Public Utility Commission for such a transfer as is required by law, or (2) compensation for the increased power-generation costs necessitated by the loss. The PUC determined that, if in fact the Idaho Power Company had failed to defend its water rights, Idaho Power customers should not be required to suffer for the company's negligence. The cost of replacing the lost power should be subtracted from the allowable company profits, shifting the cost from the rate payers to the stockholders.

Preliminary estimates of the cost involved range from $50 million per year, in a preliminary estimate by the PUC staff, down to as little as $10,000 in total. The high estimate includes the replacement (marginal or avoided) cost for *all* the power lost at all Idaho Power Company dams through failure to maintain legally entitled flows at Swan Falls Dam. The low estimate includes only a fraction of the depreciated value of the dam (which cost only $1.7 million when constructed) taking into account that the reduced flows reduce energy generation only in the low-flow seasons.

The threat of as much as a $50 million transfer from stockholders to rate payers in addition to the difficulties the power company was having in obtaining new power resources provided strong incentives for the company to act. Moreover, the Swan Falls FERC license was up for renewal, and it would be prudent to demonstrate vigorous protection of the plant's power-generation capability since renewal is by no means guaranteed. Given these incentives, the Idaho Power Company began in 1977 issuing blanket protests against all water-rights applications (except for domestic use) in the drainage area above Swan Falls Dam.

The Idaho Department of Water Resources continued to issue permits for diversions despite Idaho Power's protests, while the power company

---

[10] An additional acre-foot of water at Swan Falls could generate at most 22 kWh of electricity, whereas the same amount of water would generate 525 kWh as it passes through the middle-Snake generators.

pursued its rights through the courts. In a lawsuit popularly known as "Idaho Power Company versus the world," the power company sued irrigators and various state officials for violating its water rights. At issue in the case was the validity of the Swan Falls water rights in the face of the common assumption (at one time seemingly shared by the power company itself) that its rights were subject to subordination by a superior use—irrigation. The possibility remained that Idaho Power may have once held, but since lost, rights through forfeiture. The amount and direction of any compensation were also at issue.

The suit eventually made its way to the Idaho Supreme Court, which rendered a decision on November 19, 1982 (*Idaho Power Co.* v. *State*, 104 Idaho 575, 661 P. 2d 741 [1983]). The court overturned an earlier district court decision against the company and in favor of the irrigators, and confirmed that Idaho Power indeed has a vested property right to 8,400 cfs at Swan Falls. The state supreme court then remanded the case back to district court for determination of the forfeiture-abandonment issue, the issue of compensation, and the pending case against the Department of Water Resources and upstream irrigators.

Implications of the Swan Falls Decision

The Idaho Supreme Court decision implies either severely limited prospects for additional irrigation in southern Idaho or additional withdrawals only with the permission of, and payment of compensation to, the Idaho Power Company (and therefore implicitly the permission of the Idaho Public Utility Commission). Furthermore, some existing irrigators may be required to stop withdrawals because their water rights are invalid.

A major implication for Idaho's water institutions is that the state's water rights were determined by the state supreme court to be equally applicable to flow and diversionary uses. Priority of right is to be determined solely on the basis of the time at which the right was established, not on class of use. Hence, unsubordinated hydropower rights at dams other than Swan Falls may also constrain future diversions.

The Department of Water Resources, in its role as administrator of state water rights, was also directly affected by the decision. By implication, the department had failed to fulfill its responsibility to check for potential impacts on the power-flow rights when it granted additional irrigation-diversion rights in the upper Snake River basin. The department was attempting to manage the water resources and distribute rights so as to maintain a minimum flow at Swan Falls of only 3,500 cfs, but the water right for Idaho Power was 8,400 cfs.

Responses to the Swan Falls Decision

After assessing the initial implications of the Idaho Supreme Court's decision in the Swan Falls case, the principally affected parties—irrigators, legislators, governor's office, Department of Water Resources, Idaho Public Utility Commission, Idaho Power Company—were left to make the hard choice between trying to restore the old system, thereby risking the displeasure of electricity rate payers, or allowing the ruling to stand and alienating the traditionally powerful irrigation community. Three major alternatives were set forth: subordination of hydropower water rights, condemnation of hydropower rights, and acceptance of the court decision.

One response to the court decision was to suggest legislation declaring the flow rights for power production at Swan Falls Dam, and all other dams as well, to be subordinate to irrigation withdrawals. A main thrust of this argument is that subordination of power rights was always *intended* at all dams in Idaho. Supporters of this view point out that some of the other dams on the upper Snake River and the three dams on the middle Snake did have explicit subordination clauses in their water rights and FERC licenses. Further support for the view that subordination was intended was drawn from a 1937 U.S. Bureau of Reclamation letter (Stoutemyer, 1937), which argued that power rights should be subordinated to irrigation. Another argument favoring subordination was the state policy assumption, either implicit or explicit (as in the State Water Plan), that irrigation should be developed whenever possible and the de facto subordination policy contained in the Department of Water Resources practice of granting rights to irrigators despite prior water rights for hydropower production.

The fundamental flaw in the subordination approach was that the Idaho Supreme Court specifically addressed and rejected the argument that subordination was intended if it was not included in the rights when issued. The only way that subordination could now be incorporated into the right at Swan Falls Dam would be if the Idaho Power Company agreed to accept it. However, neither the power company nor the Public Utility Commission is likely to accept subordination without compensation.

A second proposal to deal with the supreme court decision was that the state condemn Swan Falls Dam, take over its operation, declare all existing irrigation rights valid, and resume granting water rights for new irrigation even though it reduces power production. The state certainly has the authority to condemn private property if necessary for the public good, and the State Water Plan authorizes the Department of Water Resources to operate facilities such as a hydroelectric plant. Therefore, it is technically feasible to follow this approach. However, the Idaho Power Company would have to be compensated for the value of the condemned property, and several factors complicate the determination of compensation. First, there is the question of whether compensation should include the hydropower value of water at Swan Falls only, at all downstream Idaho Power dams, or at all downstream dams. Next, the question of replacement or original cost would have to be decided. Finally, it might be difficult to convince taxpayers to approve the compensation, if it proved to be large.

A third possible response to the court decision is to accept the validity of the Swan Falls rights and proceed to determine exactly which rights of withdrawal may be invalid because of encroachment. To move in this direction, a thorough technical review of the relationship between flows and irrigation withdrawals at Swan Falls Dam would be necessary. Moreover, it almost certainly would be necessary to adjudicate all the water rights in the Snake River basin to determine which rights have illegally infringed upon the flow rights at Swan Falls.

The approach that has been undertaken by the Idaho legislature is a modification of the third alternative. A technical review is under way, and rules for validation of present withdrawal rights have been proposed. In 1983 the Idaho legislature passed Senate Bill 1180 (Chapter 5, Title 61, Idaho Code), which provides that the PUC shall have no jurisdiction over the water rights of a power company and that the PUC cannot pressure Idaho Power to defend its streamflow water rights against irrigators who have already made substantial investment based on the assumption that they had valid water rights. The senate bill also authorizes the governor to contract with Idaho Power to dismiss the suit against existing irrigators and provides that the state will hold Idaho Power blameless for rights lost by action taken if the act is later amended or repealed.

The intent of the bill was to wipe the slate clean regarding past irrigation development. There is justification in concepts of economic efficiency for such an action. To an irrigator with substantial, irrigation-specific investment, the marginal value of water made available for irrigation will be much greater than to a new irrigation development with little in-place investment. Hence, the gains from continuing to make the water available to existing irrigators may be greater than the value of the water if it were made available for power production.

A difficulty with the legislature's solution is that it appears unfair to electricity consumers. Giving up the water to the irrigators has deprived the power customers of some low-cost electricity that the courts have confirmed they were entitled to, and without offsetting compensation.

Therefore, a case could be made on equity grounds that either the water should be returned to power production or there should be compensation to the power customers.

These three alternatives and the solution that has been adopted by the legislature share a common characteristic. All are intended to deal with the immediate crisis, but none adequately addresses the general issue of reallocation of water as the value of hydropower increases relative to the value of water for irrigation. The alternatives based on condemnation or subordination contemplate continued reallocations from hydropower to irrigation but not the reverse, and the alternative based on maintaining power rights significantly inhibits the growth of irrigation. A more flexible approach would be to establish a mechanism by which water could be easily transferred between uses, once the original rights holding is established.

Idaho recently established a "water bank" institution to facilitate transfers of water between users. Idaho Power has obtained some water through this source in recent years (Hamilton and Lyman, 1983). Swan Falls rights could potentially be exchanged through the water bank for expanded irrigation—or for continued use in hydropower production.

Intrastate Allocation: Conclusion

The Swan Falls case is the first major confrontation in the Pacific Northwest between hydropower and irrigation water rights. The assumption that irrigation depletions have priority over streamflows for hydropower permeates all state water institutions in the region. In certain instances, the assumption is backed by explicit legal subordination. More generally it takes the form of administrative policy in issuing water permits. It may even be that hydropower (instream) water rights are constitutionally inferior to irrigation (diversion) rights in Oregon and Washington.

The second significant theme of the Swan Falls case is the degree of complexity surrounding the recognition of parity between hydropower and irrigation water rights. Three possible approaches to implementing parity were discussed. Each involved questions of equity (who would have to pay, or who would take a loss), as well as issues bearing on efficient allocation of water between irrigation and hydropower. These alternatives illustrate vividly the tie between rights specification and equity. On the one hand, to get on with the question of water allocation, *some* decision must be made about who has rights to what water. However, any such decision inevitably leaves both losers and winners.

The Columbia Basin Irrigation Project: State Versus Federal Control of Water

The Swan Falls case is atypical for the Pacific Northwest in that both hydropower and irrigation rights are within the state's jurisdiction. More generally, hydropower operators fall under federal jurisdiction because most major hydropower facilities in the Pacific Northwest are owned and operated by federal agencies, and FERC closely regulates the nonfederal hydropower projects. In addition, nearly half the irrigated land in the region receives water from federal projects. Therefore, the issue of how to allocate water between hydropower and irrigation uses must inevitably come to terms with both state and federal water law. In this section we highlight the federal-state jurisdictional issue, using as a case study the proposed 500,000-acre expansion of the Columbia Basin Project in central Washington.

The Columbia Basin Irrigation Project

The Columbia Basin Irrigation Project is located in the semiarid "Big Bend" area of central Washington (see map in figure 2-2). Beginning in 1918, local interests promoted the idea of building a dam at Grand Coulee, where a glacier had once blocked the Columbia River. In 1933, construction began on a version of this Grand Coulee proposal. Power generation at Grand Coulee Dam began in 1941, and water was delivered to project land beginning in 1948. The dam's power-production capacity of 6,494 megawatts is by far the largest on the Columbia-Snake river system, sufficient to supply the city of Seattle several times over. The area irrigated increased steadily to slightly over 500,000 acres by 1977, when the first half of the Columbia Basin Irrigation Project was more or less complete. The period from 1976 to 1979 saw the addition of facilities (the Second Bacon Siphon and Tunnel) that could eventually be used to serve the second half of the 1.1 million-acre project. The issue now is whether the project should be expanded to the full 1,095,000 acres authorized.

Expansion of the Columbia Basin Project would require capital investment of about $4,000 per acre in a new primary canal (the East High Canal) and associated secondary delivery and drainage systems. In the last few years, preparations to obtain the financing and approvals required for the project expansion have begun. The state, the Bureau of Reclamation, and interested private parties have jointly hired a full-time project booster and coordinator. The Bureau has formulated preliminary plans, although the major work of an environmental impact statement and detailed construction engineering have yet to begin. An attempt is being made to arrange financing through a combination of federal funds, state government participation, and a repayment contract with irrigators.

Expansion of the project would require diversion of large amounts of additional water from above Grand Coulee. Water at that point has the potential of producing 1,025 kWh per acre-foot if left to flow through turbines at eleven downstream Columbia River dams. Depletions of water to irrigate 1 acre in the project area will reduce hydropower production by about 3,600 kWh per year (Whittlesey and coauthors, 1981). In addition, irrigation pumping for expansion of the project would require approximately 2,500 kWh per year per acre irrigated.

The large impact on hydropower production and hence on regional power costs has become an issue in the debate surrounding completion of the project. Proponents point to the regional economic benefits from irrigation expansion and to a perceived national commitment to complete the project. Opponents question the economic justification for contributing large amounts of public funds to irrigate already-productive wheatland when every acre irrigated will cause more than $200 per year to be added to regional electricity-supply costs (Whittlesey, 1984). Proponents counter by pointing out that neither the water rights already held for the extension of the project nor the laws and regulations under which the Bureau of Reclamation operates require that downstream hydropower effects be taken into account in evaluation of the project.

Water Rights

Rights to water for the second half of the Columbia Basin Irrigation Project have been held in reserve since 1938. Meanwhile, downstream hydropower producers have been using the water to produce electricity.

Water Rights Held by Irrigators. Reclamation law requires that the Bureau of Reclamation obtain standard water rights under state water code for its irrigation projects. A special section of Washington water law provides that the state water agency (the Department of Ecology) can withdraw a quantity of water from appropriation, reserving it for future appropriation by the federal government. The Bureau has been issued a permit for rights to the water reserved for the first part of the Columbia Basin Project (13,450 cfs). Water for the remainder of the project (11,500 cfs) continues to be reserved with a priority date of 1938. When the Bureau is ready to irrigate, it proceeds through the standard application-permit-certificate procedure set by state water law.

While the Bureau of Reclamation appears to have secure rights to the water required to complete the project, there are two potential sources of insecurity in these rights. First, the withdrawal of rights for the second half of the project must be renewed in 1989. The renewal presents an opportunity for hydropower and other interests to offer opposing views and for the state to negotiate or terminate the water reservation if it so desires.

Second, the state must still issue a permit once the Bureau applies for the water. According to state water law, the water agency has the authority to condition or reject applications, following the criterion: "allocation of waters among potential uses and users shall be based generally on the securing of the maximum net benefits for the people of the state" (RCW 90.54.020). In practice this provision is used when preparing river basin "instream protection plans" but is not known to have been applied to evaluation of individual water permits.

The practical significance of these two legal insecurities depends on the authority of the state to condition federal (Bureau of Reclamation) use of water and the desire of the state to do so. The Supreme Court has ruled that states may, in fact, condition federally reserved water rights (for example, *U.S.* v. *New Mexico*, 438 U.S. 696 [1978]) and Bureau activities under Reclamation law (*California* v. *U.S.*, 438 U.S. 645 [1978]). At least, it has long been established that these federal uses are subject to the usual rules regarding priority of use. There are, however, at least two reasons to question the state's authority in the Columbia Basin Project case. First, state law places water claims by the Bureau of Reclamation in a special status with a priority date of 1938 regardless of when application is actually made. It may be argued that state law prohibits further substantive modification of these rights. The Department of Ecology has, for example, assumed that the Bureau's rights cannot be brought under the state's instream regulations for the Columbia even though the law specifically allows the agency to attach instream conditions in similar circumstances for private appropriators (Chapter 90.54 RCW). The presumption is that the water was withdrawn to preserve it for a specific future use; it would be contradictory to the intent of this reservation to condition it subsequently.

The second reason for questioning the state's authority to condition or deny a water right of the Columbia Basin Irrigation Project is based on the dispute over state versus federal jurisdiction. Given the supremacy of federal over state law, it might be that congressional authorization of the Columbia Basin Project gives superior status to these water rights. The Supreme Court has declared that the Federal Power Commission (now FERC) has the right to mandate water use in direct opposition to explicit state law (*First Iowa Hydro Electric Coop* v. *FPC*, 328 U.S. 151 [1946]). Moreover, the Supreme Court held in the *California* v. *U.S.* case that state laws would not hold where they conflict with congressional directives. The status of water for projects authorized under Reclamation law has not been tested.

Water Rights Held by Hydropower Operators. Although there apparently is a legal right to withdraw the water required to complete the Columbia Basin Project, the question is whether hydropower producers who already are using the water to generate electricity have vested rights with which to prevent diversion or demand compensation.

In general, water rights granted by the state to nonfederal hydropower developers do not provide protection from withdrawal by irrigators. Most state hydropower water rights are either expressly subordinated or written in terms that outline storage rights and minimum flows but do not guarantee flows. Even the exceptional cases, such as Swan Falls, where flow rights were granted, were assumed administratively to be subordinated. The subordination status of hydropower rights downstream from the Columbia River Basin Project has not been tested. In any case, the 1938 priority date of the project's water rights would give it priority over all hydropower rights at all but one of the downstream dams.

The water rights implications of the hydropower license issued by FERC are even less certain. Although FERC may issue a license in opposition to state law, current practice assumes that the FERC license grants the right to build and operate the project but does not guarantee the availability of water. FERC licenses do contain provisions governing the amounts of water to be used by the facility (for example, setting minimum flows for fishery habitat), and minimum flows set by FERC prevail over those set by the state. Whether a FERC license can be used to protect existing levels of water use is another matter, and entirely speculative.

At federal (Corps of Engineers) dams, the institutions are different, but the outcome is similar. The Corps does not apply for state licenses and permits but relies on its general mandate, specific project-authorization acts, and state concurrence for legal authority at its projects. However, the Corps' policy is to use whatever water is in the stream without questioning withdrawals even if they occur under subsequent state-granted water rights. Thus, the federal agencies follow a policy of de facto subordination of the flow rights to consumptive uses.

While the superior authority of the federal government has not been used to claim a superior right to streamflow, it seems clear that Congress has the authority to protect streamflow if it wishes to. This authority has been exercised to protect navigation (*U.S. v. Rio Grande Dam & Irrigation Co.*, 174 U.S. 690 [1899]) and Congress has completely usurped state authority to allocate the waters of the lower Colorado under the

Boulder Canyon Project Act. However, Congress has repeatedly expressed its intention to minimize the federal role in allocating water to users through numerous "saving clauses" in federal water legislation. These saving clauses provide that existing state water rights will not be changed.

Whereas protection of hydropower rights would therefore seem to require a major change in federal water law, hydropower interests might support the assertion of streamflow rights for anadromous fisheries or other instream water users as an indirect means of protecting current streamflows. In recent history, hydropower and fishery interests have for the most part been in conflict. Fishery rights have been invoked to change the seasonal pattern of streamflow to benefit fish migration while reducing hydropower production capability. These losses, however, also motivate hydropower operators to oppose more vigorously reductions in streamflow and thus provide the basis for an alliance between hydropower operators and fishery interests against water withdrawals.

The instream users might rely on the Northwest Power Planning and Conservation Act, which requires that federal agencies treat fisheries equitably in relation to other uses. The effect that withdrawals because of expansion of the Columbia Basin Project will have on fisheries must be evaluated on this criterion. In an extreme case, the Endangered Species Act might be invoked, as was considered in the late 1970s to protect streamflows for an endangered species of salmon. Finally, reserved-water and fishing rights established by Indian treaties also might serve to protect instream flows. Although these rights are still under litigation, it is clear that major federal and state action take into account the effect of the project on the Indian fishery. The Bureau of Reclamation decision to complete the Columbia Basin Project and the state's decision to issue water rights to the project certainly must be considered major actions.

Political Issues Affecting Project Completion. Although the existing federal and state water rights set the stage, it is likely that resolution of the conflict between hydropower and federal irrigation will eventually be sought through the political process. While the Columbia Basin Project is authorized in full, no funds have been appropriated for the second half. The project must be justified to Congress at a time when the public is much more critical of water projects than in the past. Moreover, budget hearings give national and regional opponents of the project an opportunity to argue their case.

Even if Congress appropriates funds toward completion of the project, it is likely to do so under terms that require state governments and local interests to share more of the cost of construction. Thus, project proponents will have to convince their own state governments to provide partial funding for the project, bringing the funding decision into a political arena where electricity customers can express their objection to the indirect costs of higher electricity rates caused by the need to replace cheap hydropower.

State Versus Federal Control of Water Allocation: Conclusion

The case of the Columbia Basin Irrigation Project is a more typical, albeit less pointed, example of the conflict between hydropower and irrigation than the Swan Falls case is, because the conflicting water rights involve both federal and state rights and laws rather than only state water rights. The basic lesson is that under current laws and administrative practices, federal or nonfederal hydropower users have little to protect them if the state grants rights for water withdrawals that reduce their generating capability. Federal irrigation in the proposed expansion of the Columbia Basin Project has a state-granted water right superior to that of other applicants by virtue of its 1938 priority date and other special provisions for federal irrigation projects. Thus, any revision of the past pattern giving all irrigation water rights de facto priority over hydropower can come only from indirect routes or from change in federal or state laws and policies.

Improving Water Allocation

The essence of the competition between irrigation and hydropower is that the marginal cost of replacing hydropower lost from river flow depletions is generally much higher than the net marginal value realized from use of the water for new irrigation developments. In the Pacific Northwest, however, water institutions generally allow expanded use of water for irrigation without compensation for the cost of lost hydropower. Consequently, irrigation development may proceed even when it reduces net economic benefits from the region's water resource.

The existing institutional structure is the legacy of a time when irrigation and other consumptive water uses were able to provide private and social benefits that clearly exceeded the small losses due to reduced streamflows. The question now is what changes can be made in the institutional structure to facilitate an efficient allocation of water under the current, greatly increased value of instream hydropower.

Approaches to Institutional Reform

The basic problem is one of allocating resources in the presence of an externality: diverters impose an external (uncompensated) cost on instream users and therefore tend to use an inefficiently large quantity of the resource. Theoretically, a problem of allocation under externalities is susceptible to state regulation or market negotiation (if rights are not ambiguous or attenuated). We adopt the market approach of assigning water rights to individual owners and allowing negotiated transfers.

If conditions are such that an ideal market can function, an efficient allocation of water among users can be obtained, regardless of the initial allocation of water rights (Coase, 1960). Thus, if the Idaho Power Company had clearly vested and transferable water rights to hydropower production in the Snake River, prospective irrigators could offer to purchase water from the power company. If the irrigators offered more than Idaho Power could expect to earn from using the water for hydropower, it would be in the interest of the company and its customers to sell rights and transfer use of the water. On the other hand, if irrigators had a vested right to water that the power company could profitably use, the water could be used to generate power, with compensation to irrigators. If value were highest in existing uses, transfer would not occur in either case. Thus, the market-transfer approach protects existing users by providing that they will have either water or compensation but provides flexibility for adjusting allocation with changes in economic circumstances.

Theoretically, a state regulator could also allocate water efficiently, but the negotiated-transfer approach is advocated here for a number of reasons. First, to allocate water, a regulatory agency would be required to calculate costs and benefits in the face of strong pressures from special interests. Second, the centralized approach presents the difficulty of adjusting to changed conditions. Rules for nonmarket allocation tend to be perpetuated long after drastic economic changes such as have recently occurred in the Northwest.

A third reason to rely on market transfer is its compatibility with existing institutions. Basically, prior-appropriation water law in the Northwest contemplates an initial assignment of property rights through a combination of self-creation and state sanction, with subsequent reallocations through private contract transfers (Trelease, 1980). Despite an increasingly important government role in overall planning and development of water use, marginal adjustments under the present system are still made by negotiation between willing sellers and buyers.

The question remains, if the current institutions provide for reallocations through transfer, why are there inefficiencies that need remedy? The problem is that current institutions fail to meet the necessary conditions for market transfers to allocate water resources efficiently. First, rights are not unambiguous and nonattenuated. Second, transfer of rights between users is not facilitated. Third, provisions have not been made for collective action where joint impact or high exclusion costs create "transaction cost" barriers to transfer. If these conditions were met, the true economic cost of the resource would be revealed by the bidding, and the resource ownership would move to its most efficient use (see, for example, Cheung, 1970; Demsetz, 1967; and Coase, 1960).

Ways in which water institutions in the Northwest fail to meet these conditions have been identified in the two case studies and the general discussion of water institutions. The following discussion of four specific avenues to the market transfer of water rights includes a brief explanation of current impediments to transfer and recommendations for reform. The four avenues are: elimination of subordination of hydropower rights, correction of interstate externalities, reconciliation between state and federal water law, and easing the transferability of water rights.

Eliminating Subordination of Hydropower Rights

The most important limit on hydropower water rights throughout the Northwest is their customary subordination (explicit or implicit) to irrigation development.[11] This attenuation, or incompleteness, in hydropower rights is a major source of the externality imposed on hydropower operators: because of subordination, water rights can be transferred from hydropower to irrigation without regard to the economic effects of the transfer. Furthermore, subordination creates a serious ambiguity in rights by denying instream users the security that consumptive users enjoy. Since the water belongs to the hydropower operator only until an irrigator decides to claim it, the power company does not know how

---

[11]The prevalence of subordination is due primarily to the development of institutions at a time when irrigation was clearly a higher-valued use. There is also an implicit and incorrect belief that consumptive uses and instream uses are qualitatively different in a way that requires different institutional treatment. Although it is true that one instream use may often be simultaneously satisfied with the same flow as one or several other instream uses, consumptive uses are incompatible with instream uses as a whole. Thus, a right to use a certain amount of streamflow for irrigation precludes use for hydropower and vice versa (as modified by the reuse of return flows).

long it will have the water. Thus, an element of uncertainty is introduced into power planning and investment programs.

Theoretically, subordination of hydropower rights could be a mechanism for granting nonattenuated water rights to irrigators *prior* to actual development. If the marginal value of water in producing power were greater than in irrigation, the hydropower users could pay prospective irrigators not to irrigate. But because of the unlimited nature of subordination, unverifiable declarations of intent to irrigate can be used to extract buy-back payments, perhaps several times for the same water. Therefore, rights must be assigned explicitly and clearly to either the hydropower operator or the irrigator if the true economic value of the water use is to be revealed.

The assignment of rights that appears to be most compatible with an efficiently functioning market and is also reasonably equitable is an allocation of full rights to current users of the water. In other words, the subordination clause should be eliminated. Present users would be, at worst, as well off as under present water use patterns since they would have the option of continuing to use water as they now do. In fact, with transferable rights and a functioning market, some present users might gain a great deal by selling or leasing their rights to other users. The only loss would be that owners of undeveloped land would no longer be able to get free water for irrigation if future technical and economic conditions are such that the owners can at least recover their costs of development.

Generally, it is within the legal authority of state water agencies to make new water permits equivalent for hydropower and irrigation. The present attenuation and ambiguity of hydropower water rights is largely a matter of policy. Even in Oregon, which defines beneficial water use in terms implying consumptive uses, the state water agency issues permits to hydropower projects as for other water uses. However, the hydropower rights usually include a subordination clause.

Elimination of subordination on existing hydropower permits is a more troublesome problem. It could be interpreted as taking away the rights to free water that prospective irrigators now enjoy. Such an interpretation could generate considerable political opposition. Also it could be difficult to establish priority among the modified permits: would the priority date of the amendments deleting subordination be the original permit date or the time of amendment? If the nonsubordination date were dated to the original permit, then water used by current irrigators might be in jeopardy, as in the Swan Falls case. There would also be a large number of issues to sort out regarding priority dates of recently added hydropower capacity and irrigation acreage throughout the region. An almost universal adjudication of the Columbia-Snake system would likely be needed to establish the amended base system of rights necessary for a system of reallocation by market transfers.

Correcting Interstate Externalities

Another externality resulting from attenuated property rights is caused by interstate water flows. Upstream states allocate flows within their borders with little regard to the effects on uses and rights in downstream states. For example, in the Swan Falls case, consideration of the impact of irrigation depletions was confined to dams in Idaho. Further, Oregon and Washington tend to compete for the essentially common-property river that flows along their common border. The problem is exacerbated because consumptive use is the only way for a state to obtain exclusive benefits for its own citizens. Some interstate interest in mutually beneficial use of the region's water resources is created by federal water resource agencies operating in all the states, by the regional power system, and by Columbia River fisheries management. However, the tendency to look only at effects within state borders is still strong.

To enhance efficiency, institutions need to be changed so that water rights issued by one state would be binding throughout the basin and transfers of rights across state lines would be recognized. Three institutional options are available: interstate compact, federal legislation (or federal agency action within present authority), and adjudication by the U.S. Supreme Court.

The Supreme Court has not issued a definitive ruling on whether states can prevent interstate water transfers. Nevertheless, the Court is the "default solution"—it will ultimately rule on interstate conflicts if issues are not resolved by other means. However, the equitable apportionment doctrine that the Court follows does not provide a predictable basis for settling interstate issues.

An interstate compact or federal legislation could provide mechanisms for resolving the interstate externality. The goal would be to achieve interstate integration by creating a single, multistate, basin-wide institutional structure for granting and transferring water rights. Each state could retain its right to issue permits within its own borders but agree to abide by a basin-wide priority system and to allow the transfer of water rights in and out of state (assuming such transfers met the usual requirements for protection of third parties).

A basin-wide water-rights structure is appealing for its potential to enhance efficiency, but states are loathe to relinquish their quasi-sovereign authority to other states or to regional collectives. Moreover, the populace generally fears water raids by neighboring states. Thus, the interstate issue may be both the key issue and the major stumbling block to reform of the current institutional structure.

Integrating Federal and State Water Law

Problems in water rights and water allocations arise partly because there are two separate water-rights laws: federal law pertaining especially to instream uses and state law pertaining especially to consumptive uses. However, instream rights cannot be protected without control over consumptive uses and vice versa. Further complications arise because various federal agencies have responsibility for different instream uses. One federal agency (U.S. Bureau of Reclamation) is primarily a consumptive user, and the states have certain responsibilities for instream uses. This bifurcated institutional structure is a major source of ambiguity in water rights. The Columbia basin case illustrates the inefficiencies that can result from unresolved conflicts between instream uses (for example, hydropower, and fish habitat) and consumptive uses (irrigation).

One avenue to bring instream and consumptive use and federal and state systems into greater harmony and to enhance efficiency could be through congressional committee action and executive oversight. Congressional committees could review broadly the implications for instream and consumptive water use before funding federal projects. Thus, funding for expansion of the Columbia Basin Irrigation Project might be questioned on the basis of including the cost of lost hydropower. Recent trends indicate greater scrutiny of such federal projects by the executive branch and by environmental groups; however, the traditional pork barrel politics of water projects would suggest little hope for thorough reviews of such projects. Furthermore, federal projects include only a part of the water allocations affected by the dichotomy between state and federal water law. The basic issue is still to clarify hydropower rights (as sanctioned by the states, FERC, and federal agencies) in relation to irrigation rights. One possibility for creation of a well-defined and transferable right to water is to require all federal projects (and federally licensed projects) to abide by general state water law. A second option would be to create a comprehensive federal water code and a federal water agency to allocate water and preside over transfers, especially between consumptive and instream uses. Given the history of water law and state sensitivities, such an institutional change is extremely unlikely. A final, more realistic approach would be to continue the present dual system of water law, but to provide explicitly a means to reconcile conflicts, such as those between hydropower rights of the Corps of Engineers and state-granted diversionary rights. What is needed is a careful inventory of federal water-related law and a set of policies and laws that systematically define the relative status of rights under federal law and state law.

Barring a purposeful legislative approach, reconciliation of state and federal law can, and ultimately will, come about only through case-by-case rulings by the U.S. Supreme Court. The problem with the judicial solution is that it is piecemeal and time consuming. Meanwhile, uncertainty will hang over all water permits and potential water transfers.

Facilitating Water Transfers

The previous three topics deal mainly with reducing the attenuation and ambiguity of water rights and thereby internalizing the full societal cost into decisions about water use. For water to migrate to higher-valued uses, however, it must also be transferable.

Even setting aside the problem of interstate transfer of water, serious institutional impediments inhibit water transfer in the Pacific Northwest. For instance, the principle of beneficial use acts as a disincentive to the transfer of water "saved" through increased efficiency in delivery. In addition, state laws often inhibit transfers of water between different classes of water use: for instance, transferring water from hydropower to irrigation or vice versa.

One issue mentioned in the context of the Swan Falls case is the need for a brokering agency or "water bank" to facilitate water transfers (Idaho Water Resources Board, 1982; Angelides and Bardach, 1978). Because of the place specificity of water diversion and streamflows, it is difficult to bring willing buyers and sellers together. A seller has a right at one place, a buyer wants to use the water in another, and the transactions costs are a barrier to transfer. A water broker or bank would allow small "deposits" and "withdrawals" to be made with much greater flexibility.

Conclusions

In summary, the following four areas are in need of institutional reform if greater efficiency is to be gained in water use in the Northwest: eliminating subordination of hydropower rights to potential future irrigation development, correcting interstate externalities, reconciling state and federal water law, and facilitating transferability of water rights. Although political realities and the nature of the water resource itself make the attainment of a perfect market impossible, institutional changes in the areas outlined would increase the social returns from the region's water resources.

Major changes in institutions and property rights imply that some water users will lose existing protection, while others will gain new rights. The political system will inevitably and properly pass judgment on the relative merits of these gains and losses. However, the sheer momentum of the current institutional structure implies that changes will be resisted and therefore will be costly to achieve even though there are substantial regional benefits to be gained. On the other hand, it is in the interest of some groups (for example, hydropower operators and irrigators with secure water rights) to pursue such changes, and institutional changes could be managed to minimize harm to existing water users.

References

Angelides, Sotirios, and Eugene Bardach. 1978. *Water Banking: How to Stop Wasting Agricultural Water* (San Francisco, Institute for Contemporary Studies).

Blumm, Michael C. 1983. "The Northwest's Hydroelectric Heritage: Prologue to the Pacific Northwest Electric Power Planning and Conservation Act," *Washington Law Review*, vol. 58, pp. 175–229.

Bonneville Power Administration. 1984. *Electric Energy Sales to Irrigators: Prospects for the Use of Nonfirm Energy* (Portland, Oreg., May).

Chaney, Ed. 1977. "The Desert Land and Carey Act in Idaho." Idaho Conservation League. Agricultural Lands Project Summary Report, June.

Cheung, S. 1970. "The Structure of a Contract and the Theory of a Non-exclusive Resource," *Journal of Law and Economics*, vol. 13, pp. 49–70.

Coase, Ronald. 1960. "Problem of Social Cost," *Journal of Law and Economics*, vol. 3, pp. 1–44.

College of Agriculture Research Center. 1981. *Demand Response to Increasing Electricity Prices by Pacific Northwest Irrigated Agriculture*. Bulletin 0897 (Pullman, Wash., Washington State University).

Demsetz, Harold. 1967. "Toward a Theory of Property Rights," *American Economic Review*, vol. 57, pp. 347–359.

Department of Agricultural Economics. 1976. *Benefits and Costs of Irrigation Development in Washington Vol. II: Final Report* (Pullman, Wash., College of Agriculture, Washington State University, October).

Hamilton, Joel R., and Ashley Lyman. 1983. *An Investigation into the Economic Impacts of Subordinating the Swan Falls Hydroelectric Water Right to Upstream Irrigation*. Idaho Water and Energy Resources Research Institute (Moscow, Idaho, University of Idaho, December).

Holmes, B. H. 1972. *A History of Federal Water Resource Programs, 1800–*

1960. U.S. Department of Agriculture, Economic Research Service. Misc. pub. no. 1223 (Washington, D.C., USDA).

———. 1979. *History of Federal Water Resource Programs and Policies, 1961–1970*. U.S. Department of Agriculture, Economic Research Service. Misc. pub. no. 1379 (Washington, D.C., USDA).

Houston, J. E., Jr. 1985. "Water and Energy Conservation Modeling in Pacific Northwest Irrigated Agriculture" (Ph.D. dissertation, Washington State University, Pullman, Washington).

Hutchins, Wells A. 1971, 1974, 1977. *Water Rights Laws in the Nineteen Western States, Three Volumes*. U.S. Department of Agriculture Economics Research Service (Washington, D.C., USDA). Idaho Water Resources Board. 1982. Idaho State Water Plan (Boise, Idaho, January).

Lee, Kai N., and Donna Lee Klemka (with Marion E. Martz). 1980. *Electric, Power and the Future of the Pacific Northwest* (Seattle, University of Washington Press).

Northwest Power Planning Council. 1983. *Northwest Conservation and Electric Power Plan, Vol. 1* (Portland, Oreg.).

Norwood, Gus. 1980. "River of Many Uses," in *Conflicts Over the Columbia River*. Seminar conducted by Water Resource Research Institute, (Corvallis, Oreg., Oregon State University, July).

———. 1981. *Columbia River Power for the People: A History of Policies of the Bonneville Power Administration*. Bonneville Power Authority (Portland, Oreg., BPA).

Office of Applied Energy Studies. 1982. *Independent Review of Washington Public Power Supply System Nuclear Plants 4 and 5: Final Report*, to the Washington State Legislature (Pullman, Wash., Washington Energy Research Center, Washington State University/University of Washington, March).

Pacific Northwest River Basins Commission (PNRBC). 1979. *Water Today and Tomorrow* (Vancouver, Wash.) vol. II.

———. 1981. "Irrigated Lands in the Pacific Northwest." Prepared by the Land Resource Committee of the PNRBC for the Depletions Task Force of the Columbia River Water Management Group.

Petke, Daniel L. 1980. "Water Quality in the Columbia River," in Water Resources Institute, *Conflict Over the Columbia River*. Proceedings of seminar conducted by the Water Resources Research Institute (Oregon State University, Corvallis, Oreg., July).

Schuy, David F. 1975. *Energy Costs of Using Columbia River for Irrigation*. Washington State University, Washington Cooperative Extension Service, E.M. 3891 (Pullman, Wash., January).

Stoutemyer, B.E. 1937. Letter to R.W. Faris, State Commissioner of Reclamation, Boise, Idaho. Available from files of Idaho Department of Water Resources, Boise, Idaho.

Trelease, Frank J. 1980. "Water Law, Policies and Politics," in *Western Water Resources*. Proceedings of a symposium sponsored by the Federal Reserve Bank of Kansas City (Boulder, Colo., Westview Press).

Utton, A. E. 1983. "The El Paso Case: Reconciling Sporhase and Vermejo," *Natural Resource Journal*, vol. 23, pp. 9–15.

Whittlesey, Norman K. 1984. "Should We Finish the Columbia Basin Project?" Paper presented at the Farm Forum, Spokane, Wash.

———, Joanne R. Buteau, Walter R. Butcher, and David Walker. 1981. *Energy Tradeoffs and Economic Feasibility of Irrigation Development in the Pacific Northwest*. Bulletin 0896 (Pullman, Wash., College of Agriculture Research Center, Washington State University).

3
Water Scarcity and Gains from Trade in Kern County, California

H. J. Vaux, Jr. *

Kern County is the third-largest county in California; its area of 8,064 square miles exceeds in size the state of Massachusetts. As figure 3-1 shows, Kern County lies astride the southern terminus of the Sierra Nevada and includes a portion of the high Mojave Desert in its eastern third. This chapter is concerned with the western third of Kern County, which encompasses 2,346 square miles of rich agricultural land at the southern end of the San Joaquin Valley. This is where the largest part of the county's economic activity occurs and where the two largest industries, petroleum and agriculture, are centered. It is also the location of the county seat of Bakersfield, the largest urban area in the county, with a 1980 population of 98,400 (State of California, 1980).

In 1980, the gross agricultural production in the valley portion of Kern County was $1.27 billion, putting Kern second only to Fresno County as an agricultural producer in California. The prominent economic position of agriculture in Kern County is attributable to the plentiful and relatively cheap supply of irrigation water. Long-term precipitation averages only between 5 and 8 inches annually, and nearly all of this falls during the winter months; there is almost no rainfall during the long hot growing season when the area's productive potential is greatest. Without water for irrigation, agriculture in Kern County would likely be limited to small dryland farms producing grain crops mainly in the winter months.

*Department of Soil and Environmental Sciences, University of California, Riverside.

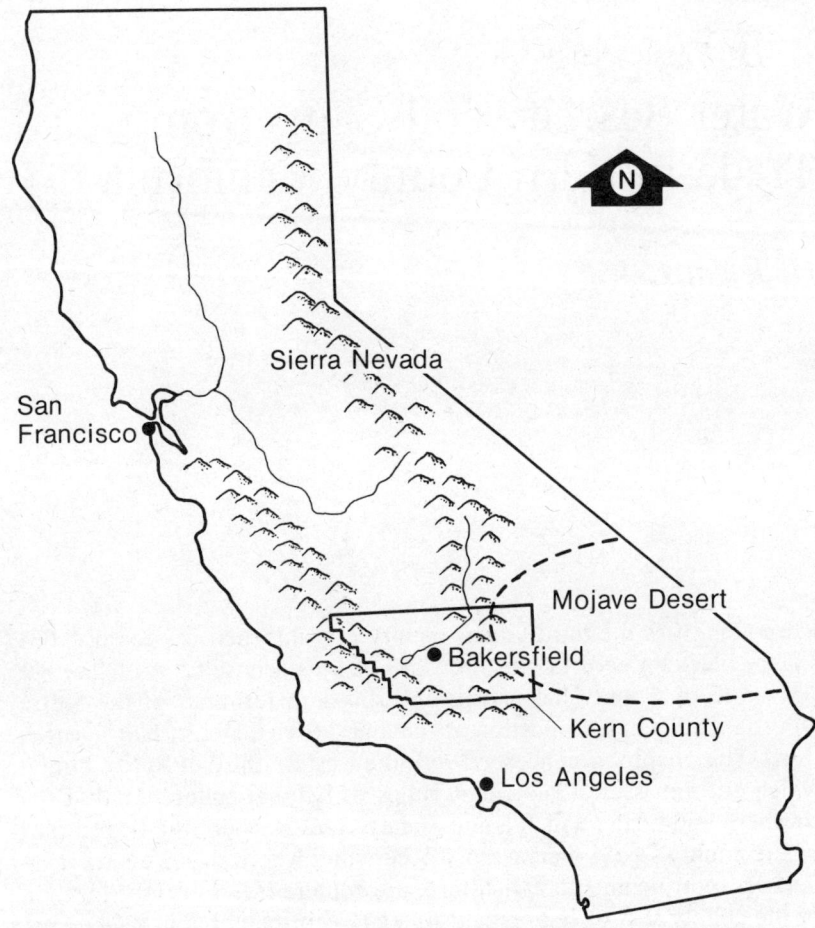

Figure 3-1. Kern County, California

Kern County Water History

Local Water Sources

The rise of irrigated agriculture in Kern County was tied to the availability of water from local sources. Early agriculture was dependent upon the flows of the Kern River, the region's largest and most significant local source of surface water. As figure 3-2 shows, the Kern River rises in the southern Sierra in the area immediately west of Mt. Whitney. From there it flows in a generally southerly direction for approximately

Figure 3-2. Streams and conveyance facilities serving Kern County

100 miles, after which it turns west to the San Joaquin Valley, entering at a point immediately east of Bakersfield. In its natural state, the river ultimately crossed the valley and ponded in Lake Buena Vista and, in wet years, several other lakes, all of which were intermittently dry. The Kern River basin, then, is a closed basin with no outlet to the sea except in years of exceptionally heavy rainfall, when the terminal lakes overflow and channels are established to the San Joaquin River.

In the nineteenth century, irrigated agriculture in the region was limited to the lands immediately adjacent to the Kern River and several minor streams. This early irrigation was affected by substantial yearly variation in Kern River flows. Over the period of record (1894–1982), annual unregulated Kern River flows averaged 705,000 acre-feet, measured near Bakersfield. However, unregulated annual flows have ranged from a minimum of 177,000 acre-feet in 1961 to a maximum of 2.3 million acre-feet in 1969 (Kern County Water Agency, 1983). Equally important is the seasonal variability in flows, which ranges, on the average, from 1 percent of annual runoff in November to 33 percent in June. Less than 20 percent of the runoff occurs during the driest months of July, August, and September.

The substantial aquifers of Kern County were not tapped until late in the nineteenth century, when pumps first became available. Initially, groundwater extraction was constrained by pumping technology. As technology advanced, more areas were brought into production. By the 1930s the advent of increasingly sophisticated pumps and well-drilling technology had led to a serious overdraft problem, which resulted in efforts to import surface supplies. This began the modern pattern of irrigated agriculture in Kern County, which has been characterized by recurring cycles of groundwater depletion halted through importing supplemental surface supplies, followed by a short lag and then by a resumption of groundwater depletion.

Imported water supplies were first made available through the Central Valley Project. This project was originally envisaged as a state undertaking but fell victim to the Depression, at which point the Bureau of Reclamation accepted the task of making it a reality. The part of the Central Valley Project (CVP) that is relevant here includes the Friant Dam on the San Joaquin River and the Friant-Kern Canal, which runs 150 miles south from Friant Dam to a point in Kern County just south of Bakersfield, as illustrated in figure 3-2. It brought surface water to areas on the east side of the valley with rapidly declining groundwater tables and to areas where surface water supplies had been unreliable. Over the last twenty years, the Friant-Kern Canal has conveyed to Kern County an average of 332,400 acre-feet annually.

In 1954, the Corps of Engineers completed Isabella Dam on the Kern River. Although this project was designed primarily for flood control, it also regulated Kern River flows and made the river a more reliable source of irrigation water. The dam minimized flows to Lake Buena Vista in the spring, thereby conserving them for use in irrigation during the summer months. The combination of water supplies from the Central Valley Project and the regulation of Kern River flows abated the decline in groundwater tables only temporarily, however.

By the late 1950s, overdraft was estimated to be more than 700,000 acre-feet annually. Agricultural interests sought new sources of imported supplies, this time from the state of California. The appeal was answered in 1961 by the creation of the Kern County Water Agency, which was granted authority to acquire water supplies for Kern County and to participate in flood control, drainage, and reclamation activities. Of principal interest, however, is the fact that the Kern County Water Agency is the contracting entity for water from the State Water Project (SWP). In the mid-1960s, the agency contracted with the state for an ultimate annual entitlement of 1.253 million acre-feet (maf). Deliveries, which were begun in 1968, were to be in accord with a buildup in entitlements that would reach the ultimate level by 1995. In 1980, the entitlements totaled 919,200 acre-feet. The SWP water is delivered to Kern County from the delta of the Sacramento and San Joaquin Rivers via the California Aqueduct. As shown in figure 3-2, the aqueduct lies on the western side of the San Joaquin Valley. Ultimately, it traverses the Tehachapi Mountains and enters the Los Angeles basin, a fact that significantly influences the quantity of water that Kern County actually receives.

Although Kern County obtains water from four distinct sources (local surface water, local groundwater, the federal Central Valley Project, and the State Water Project), the quality of its water supplies is relatively constant. The maximum total dissolved solids in any of these supplies is 400 ppm. As a consequence, the quality of water supplies poses no special problems even though those supplies come from disparate sources. Kern County, however, is faced with a water quality problem that stems from the fact that it lies in a closed basin. Because there is no regular pathway through which drainage water can be expelled from the basin, salinization, together with the presence of highly saline perched groundwater poses a threat to the productivity of perhaps 40,000 acres of land. Although desalinization and irrigation of this land will ultimately be related, at present the drainage problem is perceived to be quite distinct from the problem of obtaining adequate water supplies and is thus beyond the scope of this chapter.

The irrigation picture of Kern County that emerges for 1980 is both impressive and problematic. The valley portion of the county contains 944,000 irrigated acres that produce more than a billion dollars in gross revenues annually. Cotton, vegetables, and perennial crops are the most important in terms of both acreage and revenues as shown in table 3-1. Irrigation is accomplished, on the average, with 3.28 million acre-feet of water annually. An additional 110,000 acre-feet are used for municipal and industrial purposes. Table 3-2 shows that, of the amount of water devoted to agriculture, 58 percent is from local sources, including the Kern River and groundwater pumping, while the remaining 42 percent is imported via federal (13 percent) and state (29 percent) facilities from other basins.

TABLE 3-1. Acreage and Gross Value of Irrigated Crops in Kern County, 1980

Crop	Acres	Gross value ($1,000)
Cotton	396,480	354,427
Grapes	84,960	194,640
Vegetables	84,960	155,333
Almonds	66,080	147,237
Citrus	28,320	58,506
Alfalfa	94,400	58,422
Grain	103,840	27,013
Other	84,960	210,407
Total	944,000	1,205,985

Source: Kern County Agricultural Commissioner, *Agricultural Crop Report*, 1981, Kern County (Bakersfield, Calif.).

Because of the continuing expansion of irrigated agriculture, Kern County is once again faced with groundwater overdraft. Current estimates suggest that in an average year the prevailing pattern of agricultural development requires approximately 370,000 acre-feet of overdraft (Associated Engineering Consultants, 1983). The problem is intensified by falling groundwater tables throughout the portion of the county that must rely solely on groundwater for its irrigation supplies. Recent and prospective increases in the cost of energy suggest that some decline in irrigated acreage is inevitable unless there are either additional supplies or changes in the patterns of water use and management.

Immediate prospects for additional supplies from external surface sources are not bright, even though the Kern County Water Agency has contracted for increasing entitlements from the state through 1990. It is unlikely that the state will be able to deliver substantially more than 900,000 acre-feet of the ultimate 1.25 maf total before the year 2000

TABLE 3-2. Sources and Uses of Water in Kern County, 1980 (1,000 acre-feet)

	Agriculture	Municipal/industrial
Kern River	610	—
Central Valley Project (federal)	420	—
State Water Project	960	40
Groundwater	1,290	70
Total	3,280	110

Source: Associated Engineering Consultants, *Report on Investigation of Optimization and Enhancement of the Water Supplies of Kern County* (Bakersfield, Calif.) January 1983.

because of political impediments to construction of a cross-delta facility needed to boost deliveries to the full entitlement level. Kern County's problem is further intensified by the fact that current levels of water use are sustained, in part, by deliveries of so-called surplus water from the State Water Project. This surplus water is water that has been contracted for by agencies south of the Tehachapis that are not currently taking their full entitlements, leaving some water available to Kern County and other Central Valley users for only the cost of conveying it. When these southern contractors begin to take their entitlements in response to the Supreme Court-mandated reduction in California's share of Colorado River waters, the surplus available to Central Valley users will be restricted, perhaps sharply. This means that new supplies will be required simply to maintain current levels of water use in Kern County.

The water situation in Kern County mirrors in a general way the water situation faced by the entire state of California. Over the last hundred years, substantial public investment in water storage and conveyance facilities has succeeded in moving water from areas of the state where it is plentiful to the major urban centers and most productive agricultural regions. In general, water scarcity has been ameliorated by developing additional supplies, and little or no attention has been given to the possibilities for regulating demand. But now the supply augmentation strategy that has served Kern County and the state well for the last hundred years is becoming less viable.

The cheaper, easy-to-exploit water supplies have already been developed. With few exceptions, new sources of supply will be much more costly to develop. In addition, the relative increases in construction costs and the cost of energy needed to convey new supplies have risen. Many users and potential users argue that they will be unable to defray the full costs of new development, which suggests the need for further public subsidy. Yet increasing competition for scarce public dollars—coupled with political pressures to restrain public spending—suggest that such subsidies may not be as readily forthcoming as in the past. Moreover, pressures to preserve the environmental values associated with free-flowing streams have further frustrated efforts to develop new supplies (Howitt and coauthors, 1982).

The problem in Kern County is perhaps more acute than in most other parts of the state. In the state as a whole, the need for new supplies is perceived as arising from the continued growth in population, a familiar story. In Kern County, growth is a consideration, but preserving current levels of use is of far more concern. A report commissioned by the Kern County Water Agency estimates that activities supported by groundwater overdraft generate $24 million of value-added each year. The principal worry, then, centers on the economic dislocation that would occur when overdrafted aquifers reach equilibrium and activities now dependent on overdraft can no longer be supported (Northwest Economic Associates, 1983).

A substantial literature records the inefficiency of water use in California. Hirshleifer and coauthors (1960) identified the conditions required for efficient investment in water resource facilities and documented some of the inefficiencies inherent in the development of California's water system. Bain and coauthors (1966) comprehensively examined the northern California water industry and concluded that the institutions governing that industry create misallocations of water. They identified, moreover, a number of factors that tend to impede the exchange of water and the establishment of water markets and documented the fact that water in California is locked into many inefficient uses. In general, new uses must be served by new sources of supply since there are no mechanisms through which water is easily transferred from existing, low-value uses to new high-value uses.

More recently, Vaux and Howitt (1984) have shown that rationalization of water pricing policies and establishment of limited quasi-markets for water could enable Californians to adjust to water scarcity with minimal new water development. The annual gains from trade in such markets would amount to more than $70 million now and would grow substantially over time. This work suggests that removing even some of the inefficiencies inherent in prevailing patterns of water use in California would both benefit the trading parties and provide water users more accurate signals on the scarcity of water in California's semiarid environment.

There is no direct evidence to indicate how much Kern County might benefit from market-like institutions that facilitate trade. It can be argued inferentially from existing information that markets would result in more efficient use of water in Kern County and could make additional supplies of water available for agriculture. Whether such markets can be developed is another question, however, since there are a host of institutional barriers to market exchanges of water.

To understand both the potential for water trade and the barriers to it, it is necessary to know something about the institutions that govern water supply and use activities in Kern County. These institutions are discussed in the next section. In the third section, arguments in support of potential gains from trade are advanced, and in the fourth section, various impediments to trade are identified and analyzed. A final section presents concluding comments.

Institutions Governing Water Use and Supply

Kern County is characteristic of California in that it possesses a bewildering array of water institutions. The discussion here is restricted to four broad classes of institutions that shape the allocation of water in Kern County and that will influence critically the extent of market-type transfers, if any. The institutions considered are water rights, purveying agencies, existing exchange arrangements, and pricing and allocation rules.

Water Rights

California is unique in its dual system of water rights, in which both riparian and appropriative rights exist side by side. (This discussion of water rights is taken largely from Hutchins (1956) and the 1978 *Final Report* of the Governor's Commission to Review California Water Rights Law.) Riparian rights to surface water are use rights attached to any tract of land adjoining a stream or lake. A riparian right is limited to the quantity of water that can be put to "reasonable and beneficial" use on the tract of land in question. The right cannot be expanded by adding dry land to the tract to which the right accrues, but it can be diminished if the tract is subdivided and the size of the parcel that abuts the stream or lake reduced. Where the sum of riparian rights exceeds the common water supply, riparians are obliged to share the water equitably. Riparian rights are held almost exclusively by private individuals, since public agencies and private companies cannot acquire them except for use on lands to which they are appurtenant.

Two features of riparian rights are especially significant. First, riparian rights in virtually all instances take precedence over appropriative rights. That is, a typical appropriative right can be exercised only on water that is surplus to water subject to riparian rights. In time of shortfall, riparian rights are superior to all appropriative rights except for the very few that were established prior to the riparian right (Bowden and coauthors, 1982). Second, in the strict sense riparian rights may not be transferred independently of the land to which they attach. The water may be captured through adverse possession, but in such instances the riparian right is not transferred; instead, a prescriptive right, similar to an appropriative right, is established in its place. It is thus clear that under the prevailing legal doctrine in California a riparian right cannot be traded except in association with the land to which it attaches.

Appropriative rights to surface waters are obtained by using the water continuously and for a reasonable, beneficial purpose. Failure to use the water for a period of five years results in the loss of the right. Appropriative rights can be established only for water that is "surplus" to the reasonable and beneficial requirement for other rights, both riparian and senior appropriative rights, established in the water source. Seniority in appropriative rights is established by the first date of appropriation, with an early right being senior to a later one. Today, appropriative rights are licensed by filing with the State Water Resources Control Board, which then holds hearings and presents findings with respect to the availability of surplus water, the absence of injury to third parties with standing, and the "reasonableness" of the appropriation. The licensing process has been in effect only since 1914, and rights established prior to that date have not been recorded.

Appropriative rights may be established by private users or by intermediaries such as local public agencies or private water companies. In the latter case, the members of the intermediary are viewed as having a beneficial interest in the water. Thus, appropriative rights to surface water tend to reside ultimately in the users of the water (Bain and coauthors, 1966). The appropriative right is not inherently tied to the land, and the site of use or point of diversion may be changed subject to the approval of the State Water Resources Control Board.

However, there is a tendency for appropriative rights to become appurtenant to the land on which they are used and which are specified on the license issued by the state. Although it is usually a simple matter to change the kind of use—from irrigation to municipal and industrial use, for example—changing the location of use is more problematical. The law recognizes the interrelatedness of uses, and changes in the site of use or point of diversion cannot occur without approval from the Water Resources Control Board. This approval is not granted if injury will result to third-party users. The board also recognizes that changes in the use site do not generally eliminate all demands for water for the lands from which the water has been transferred. The result is an institutional bias against trade, created by fears that water trading may simply be an indirect way of generating demands for new supplies. This bias, coupled with the fact that an appropriative right nominally passes from buyer to seller with the sale of land, results in a strong tendency for appropriative rights to be appurtenant to the land. However, in a strict legal sense such rights may be separated from the land and thus are not tied to the land to the same extent as riparian rights.

Groundwater is also subject to a dual system of water rights. The counterpart of the riparian right is the correlative right, which may be established by users overlying the groundwater formation. Correlative rights entitle holders to make reasonable and beneficial use of groundwater on overlying lands and oblige them to share equitably in the event that all correlative uses cannot be fully served. Correlative rights cannot be established for lands not overlying the aquifer. For nonoverlying lands, appropriative rights to groundwater—similar in most respects to appropriative rights to surface water—may be established. A critical difference is that there is no requirement for filing and licensing to establish an appropriative groundwater right; it is necessary only to initiate use and ensure that it is continuous. The result is that groundwater rights are not recorded or quantified except in a few urban basins where there has been extensive litigation. Appropriative rights to groundwater are inferior to correlative rights.

In Kern County, as in most other agricultural areas of California, the permissiveness of groundwater law has fostered a situation in which there are virtually no restrictions on groundwater pumping other than the economic restrictions imposed by cost. The uncertainty associated with the failure to quantify groundwater rights also extends to some classes of surface water rights. Riparian rights are probably the most uncertain because they are subject to substantial variation in quantity (Lee, 1977). Appropriative rights, however, are also vague as to amount. The pre-1914 rights have never been quantified, and those established between 1914 and 1969 were recorded only in terms of flow rates and seasonal restrictions and not total quantity (Howitt and coauthors, 1982). The unrecorded pre-1914 rights may be especially significant. Estimates suggest that they may amount to as much as 8 maf statewide (Governor's Commission to Review California Water Rights Law, 1978).

This brief overview suggests that California's system of water rights is complex, inflexible, and fraught with uncertainties. The specific ways in which water law may impede the establishment of market-like institutions will be discussed further in the fourth section.

Water Agencies

It is a striking characteristic of the Central Valley region of Kern County that well over 90 percent of the irrigable land falls within the boundaries of some special water districts formed to acquire and purvey water to local users. The districts are overlain, in turn, by the countywide Kern County Water Agency, whose members are a subset of the special districts. The location of the districts is illustrated in figure 3-3. These districts, which Bain and coauthors (1966) characterize as "user cooperatives," have been formed under a variety of provisions in the California Water Code for the general purposes of acquiring, storing, distributing, and conserving water. There are some 138 different types of water districts (California Department of Water Resources, 1978), but in Kern County, the predominant forms are the California Water District

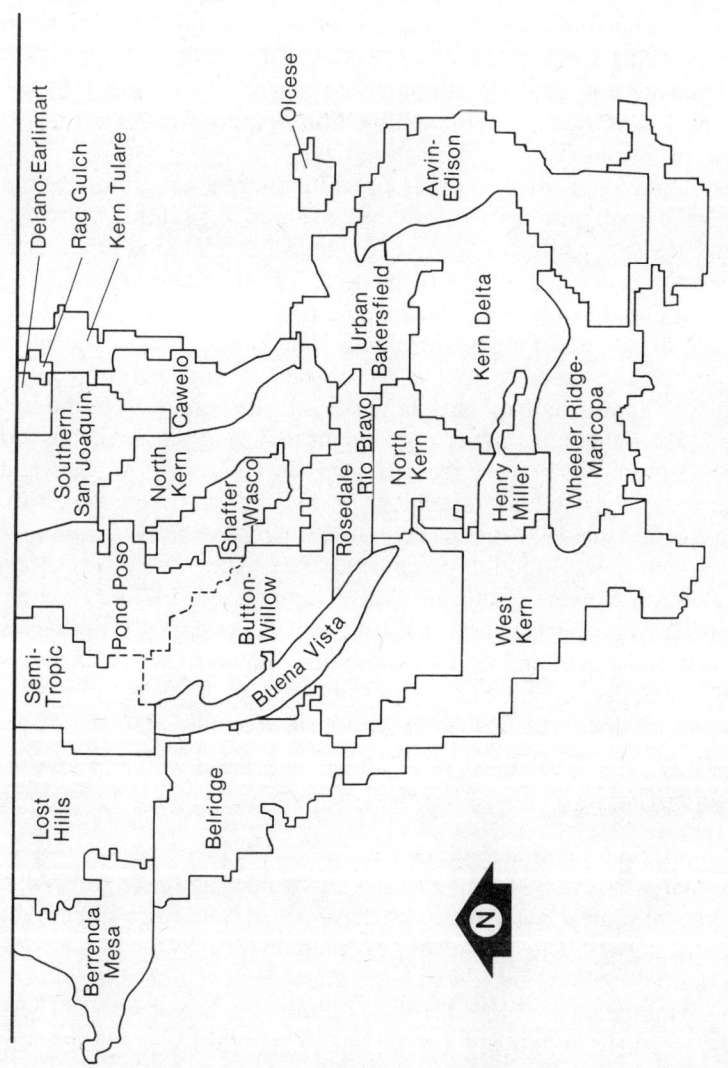

Figure 3-3. Irrigation districts of Kern County

(9) and the Water Storage District (7). In addition, there are two irrigation districts and one municipal utility district.

It is difficult to generalize about the characteristics of these types of districts. Indeed, the history of each district is distinctive and often complicated. Under the law, all districts are permitted to tax, contract with state and federal agencies, issue bonds, and receive revenues. All districts are explicitly constrained from making a profit (Phelps and coauthors, 1978). The details of taxing powers, the makeup of the governing structure, restrictions on the area to be served, and the extent of regulation by the state government differ by district type. These distinctions, however, have an insignificant effect on the ability of various districts to acquire, manage, or trade water.

The user-cooperative nature of the districts has arisen because of the potential for scale economies; districts can acquire and distribute surface water far more cheaply than individual water users. Nearly all of the districts have access to both surface and groundwater supplies and some have multiple sources of surface supply. Normally, groundwater is pumped by individual growers, although several districts operate their own pumps. In Table 3-3, each district is identified according to whether the major source of its surface supplies is local or from the CVP or the SWP. Among the districts listed, only the North Kern Water Storage District has a purely local source of supply. A majority of the districts have entitlements to water from the State Water Project and are, therefore, member districts of the Kern County Water Agency.

The Kern County Water Agency is a distinct type of special district. The agency overlays the entire county, although the only districts that

TABLE 3-3. Kern County Water Districts Listed by Major Source of Surface Supply

Kern River entitlements	Friant-Kern (Central Valley Project)	State Water Project
Buena Vista	Arvin-Edison	Belridge
Kern Delta	Delano-Earlimart	Berrenda Mesa
North Kern	Rag Gulch	Buttonwillow
	Shafter-Wasco	Cawelo
	Southern San Joaquin	Henry Miller
		Lost Hills
		Pond Poso
		Rosedale-Rio Bravo
		Semitropic
		West Kern
		Wheeler Ridge

Source: Adapted from William D. Watson, Carole Frank Nuckton, and Richard E. Howitt, *Crop Production and Water Supply Characteristics of Kern County*, Bulletin 1895 (Davis, Calif., Giannini Foundation) April 1980.

are members are those with entitlements to state water. It was formed for the purpose of contracting with the state of California for water from the State Water Project and has access to the entire tax base of Kern County for help in financing its operations. Its primary function is to sell water from the State Water Project to the member districts at wholesale rates. It also has responsibilities for flood control, reclamation, water storage, and a variety of other water-related activities. Together, the individual member districts and the Kern County Water Agency determine the allocation of all waters imported from the State Water Project. The districts that receive water from the CVP interact directly with the Bureau of Reclamation.

Water districts, through their contracting and financing powers, as well as their internal allocative decisions, exert a predominant influence on the way water is managed in Kern County. Virtually the only waters unaffected by the districts are surface waters to which there are private riparian and appropriative rights. Although the districts have no explicit statutory or administrative authority with respect to groundwater management, the availability of groundwater is critically influenced by the way surface waters are managed. Specifically, policies that require that surface water prices be set at levels below the cost of groundwater pumping induce users to substitute surface water for groundwater in joint supply areas. In addition, some districts recharge groundwater directly. These measures depend on the availability of surface water supplies, which may be augmented through physical and contractual exchange arrangements.

Exchange Arrangements

In Kern County there are three major kinds of exchange facilities that permit the exchange or substitution of water from different sources. The surface water exchange facilities include the Kern River Intertie and the Cross Valley Canal. The location of these facilities is shown in figure 3-4. The Kern River-California Aqueduct Intertie is a physical facility that enables water flowing in the lower reaches of the Kern River to be put into the California Aqueduct and made available to other users along the aqueduct. The intertie, which was constructed under the auspices of the Army Corps of Engineers for flood control, allows for two different types of exchange. Both have been realized in recent years.

First, in years of high flows in the Kern River, excess river water can be placed directly into the California Aqueduct. In 1978, 1980, and 1982, a total of nearly 318,000 acre-feet were discharged into the aqueduct (Kern County Water Agency, 1983). This water was used by downstream contractors both in Kern County and south of the Tehachapis,

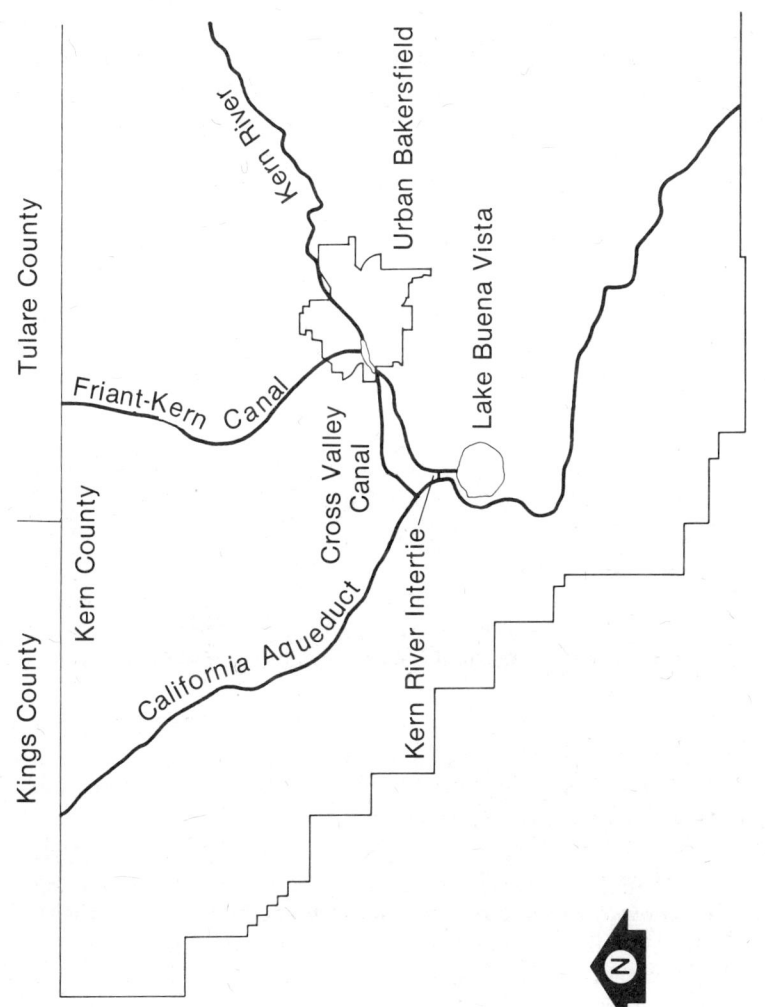

Figure 3-4. Major water conveyance facilities in Kern County

thereby reducing the quantity of water that the state would otherwise have had to deliver in its system. In the same years, the second type of exchange possibility was also utilized. Water that is surplus to the demands of the Friant-Kern water users, as well as flood flows from the Kings and Kaweah rivers, can be delivered to the Kern River via the Friant-Kern Canal and ultimately to the aqueduct through the intertie. In the three years that the intertie has been operating, a little more than 21,000 acre-feet from the Friant-Kern Canal has entered the California Aqueduct through the intertie. The intertie, then, effectively joins the Kern River and the Friant-Kern Canal to the California Aqueduct and permits excess water along the Friant-Kern Canal or in the Kern River to be made available to state contractors.

The Cross Valley Canal is a second major facility that enhances the flexibility of surface water allocations. The canal was financed by five of the agricultural districts and the Improvement District that serves the urban Bakersfield area. It permits state water to be transported from the California Aqueduct 20 miles eastward and raised 80 feet, enabling delivery of state water to Bakersfield and districts on the east side of the valley. Some 280,000 acre-feet per year was anticipated, but experience in 1979, when 350,000 acre-feet was delivered, showed that the canal could carry considerably more (Kern County Water Agency, 1982). The canal allows water to be delivered directly to the Arvin-Edison Water Storage District and, through an exchange arrangement, to the Cawelo, Rag Gulch, and Kern-Tulare water districts as well as a number of districts outside of Kern County. In addition, the Rosedale-Rio Bravo and Arvin-Edison water storage districts and the Kern County Water Agency utilize water from the canal in groundwater spreading programs. Thus the Kern River Intertie and the Cross Valley Canal permit large physical exchanges of surface water and provide substantial capacity to recharge the county's aquifers with the surplus.

Last, Kern County's physical exchange facilities include the spreading grounds operated for direct groundwater recharge by several of the districts and the Kern County Water Agency. The condition of Kern County's groundwater basins is of concern for two reasons. First, approximately 39 percent of the net irrigated acreage in the valley portion of the county is completely dependent on groundwater. These areas are vulnerable to depletion by groundwater mining. Second, the erratic flows of the Kern River and other rivers whose supplies are imported into Kern County mean that in any given year surface supplies can be extremely short. For example, in the drought year of 1977, total surface deliveries from all sources were reduced to 40 percent of what would have been expected in a normal year. The ability of the majority of the growers in the county to pump groundwater substantially lessened the economic impact of the drought.

As a result, the Kern County Water Agency and certain water districts have undertaken conjunctive use programs in which surplus water available in some years is applied to spreading grounds and stored in groundwater basins for use in dry years. The Kern County Water Agency estimates that nearly 2.5 maf were lost from groundwater storage during the 1976/77 drought, but since that time there have been net additions of over 1.9 maf from spreading and percolation in unlined canals (Kern County Water Agency, 1983). Formal water-spreading facilities include a 2,800-acre facility operated by the city of Bakersfield and somewhat smaller facilities operated by the Rosedale-Rio Bravo, Arvin-Edison, and North Kern water storage districts (Associated Engineering Consultants, 1983).

The final element in the exchange arrangements available to Kern County is contractual. To understand what is involved, it is necessary to distinguish between entitlement or firm water and surplus water. Contractors with both the CVP and the State Water Project contract for some quantity of entitlement or firm water. The existence of climatic variations means that entitlement water usually amounts to less than is available in all but dry years. In wet years, surplus water is available and is usually sold at low prices. In Kern County there is another significant source of surplus water. The contractors south of the Tehachapis in the southern urban regions contracted for entitlement water well in advance of the need. As a result, they have been willing to make much of this water available as surplus to districts in Kern County in order to avoid the substantial energy costs that must be incurred to lift the water out of the Central Valley and over the Tehachapis. As noted earlier, when these entitlements are fully accepted, this significant source of surplus water will disappear.

The fact that the State Water Project provides supplies to the Metropolitan Water District of Southern California, whose main source of supply is the Colorado River, means that hydrologic events on the Colorado River can influence water supply conditions in Kern County. In anticipation of a continuing drought in 1978, the Kern County Water Agency consummated an exchange agreement with the Metropolitan Water District that provided for the agency to pay the Metropolitan Water District to pump water from the Colorado River in exchange for a portion of the district's entitlement for that year. In 1978, the Metropolitan Water District pumped 30,000 acre-feet from the Colorado River at Kern County Water Agency expense and stored it in groundwater basins in the southern part of the state. The agency received, in turn, a claim to some 25,000 acre-feet of Metropolitan's State Water Project entitlement in some future dry year. This illustrates one of the many possibilities that the interrelatedness of Kern County's surface supply facilities offer. The trading possibilities between the Imperial Irrigation District and the Metropolitan Water District discussed in chapter 4 could have a significant impact on the availability of surplus surface water in Kern County.

These physical and contractual exchange arrangements provide Kern County with a variety of possibilities for adjusting to shortfalls in one or more of its regular sources of supply—providing that there is surplus water available somewhere in the substantial plumbing system that serves Kern County. The extent to which such possibilities can be realized is well illustrated by the case of the Arvin-Edison Water Storage District. Historically, the area served by this district relied primarily on groundwater, supplemented by the small and erratic flows of some minor local streams. One result was the progressive decline in groundwater depths from an average level of 250 feet in 1950 to more than 450 feet by 1965. With the completion of the Friant-Kern Canal and the district's internal distribution facilities, Arvin-Edison's contract for CVP water included 40,000 acre-feet of firm or Class 1 supply and 313,000 acre-feet of surplus or Class 2 supply. The district allocated this surface water to specific areas where groundwater tables were especially low and used the remainder in district-owned spreading works to recharge the groundwater.

Yet Arvin still faced a problem. Only a small portion of its total CVP supply was firm, making the district subject to substantial shortfalls in surface deliveries in drier years. The need was not so much for more water as it was for more guaranteed water. In the mid-1970s, the district was able to effectuate a solution in the form of a large water exchange. The exchange required Arvin-Edison to make the first 174,300 acre-feet of its CVP entitlement available to some nine irrigation districts upstream on the Friant-Kern Canal. In return, Arvin received 128,300 acre-feet of firm supply, delivered by the Bureau of Reclamation from the Sacramento-San Joaquin Delta (where it was essentially surplus to other needs) through the California Aqueduct and the Cross-Valley Canal (Arvin-Edison Water Storage District, 1983).

It is interesting that the terms of trade required Arvin to give up more water than it received, since the average delivery of its original Class 1 and Class 2 water was 128,300 acre-feet. The financial circumstances of the exchange are also of interest. Arvin-Edison pays a portion of the annualized capital cost of the Cross Valley Canal and continues to pay for its total CVP surplus and entitlement water. The Friant-Kern Exchange Districts pay the federal water charges for the "new water," plus the costs of conveying it through the California Aqueduct and Cross

Valley Canal and a significant portion of the capital costs of the Cross Valley Canal. These costs total approximately $37 per acre-foot. The resulting exchange enables Arvin to increase its commitment to deliver firm water by paying its share of the capital costs of the Cross Valley Canal, which are relatively small. The exchange also enhances the groundwater recharge program because increases in firm surface water supplies reduce substantially the extent to which groundwater pumping must be relied upon in both normal and dry years (Arvin-Edison Water Storage District, 1983).

A second long-term exchange agreement involves the Buena Vista Water Storage District and the Kern Delta Water District. Under this agreement, a portion of Buena Vista's Kern River entitlement water is delivered to Kern Delta. In exchange, Buena Vista receives an equal amount of Kern Delta's state water entitlement. The long-term average quantity of water to be exchanged is 22,000 acre-feet, significantly less than the quantities involved in the Arvin-Edison exchange. This exchange permits the Kern Delta Water District to obtain water for which it has contracted with the state even though it has no means of turning water out of the state aqueduct. Buena Vista has direct access to the aqueduct (Associated Engineering Consultants, 1983).

In addition to these two long-term exchange arrangements, there are a number of short-term, year-to-year arrangements that permit districts to exchange water from different sources. All of these exchange arrangements indicate the ability and willingness of the Kern County districts to engage in trade that is mutually beneficial. At the same time, all exchanges involve in-kind swapping that results in either modest increases in total quantities of available water or augmentation of firm supplies. There are no exchange agreements in which one of the parties ends up with less firm water than it had prior to the exchange. To date, there has been no internal exchange in which a district or group of users has received monetary compensation for reducing its water use, although in some cases districts have been able to sell surplus water in wet years.

Pricing and Allocation Rules

In Kern County, as in all of California, the price of water is exclusively related to the financial costs of supplying it. That is, the price is established so as to defray both the capital and variable costs of delivering the water from its place of origin to the site of use. The existence of federal subsidies and the practice of melding costs and charging average rather than marginal costs means that prices are often lower than the economic costs. Moreover, no opportunity cost or value is imputed to the water itself, which is thus treated as a free resource. The basic structure of prices at the retail level is generally consistent with the classical form of utility pricing, in which there is both a fixed charge and a variable charge.

The generalized form of the pricing structure found in the various districts in Kern County involves a variable charge and two kinds of fixed charges. The variable cost or user charge is assessed by the district on each acre-foot delivered to the farm gate. Commonly, though not always, this charge reflects the variable cost of delivering the water. In districts where there are significant elevation changes, the charge will vary depending upon the elevation to which the water must be lifted. The fixed costs consist first of a standby or water availability charge imposed on a per-acre basis, and, second, of a component imposed as a property tax. Revenues derived from the fixed costs are generally used to service district debts for distribution facilities and pay maintenance and administrative costs. In some instances, the fixed costs are set high enough to cover a portion of the variable costs, providing districts with the relatively stable levels of revenues needed to cover their fixed obligations. Gardner (1983) has suggested that this practice is employed to insulate district revenues from shortfalls attributable to reductions in water use caused by increases in the variable costs.

The prices actually charged by retailing agencies also reflect average cost pricing. Although this practice may have begun in response to the decreasing-cost tendencies of water purveying activities, it has persisted despite indications that increasing competition for water, along with escalating energy prices, may have changed the picture to one of increasing costs. The persistence of average cost pricing stems, first, from user pressures to keep all prices at "affordable" levels, and second, from legal provisions that constrain districts from making profits (Phelps and coauthors, 1978). The presence of surplus water that is usually sold at wholesale prices sufficient to cover only the variable costs of conveyance, and that is therefore substantially cheaper than firm water, is a contributing factor. Districts are compelled to cover the costs of their firm water entitlements first. By melding the cheaper costs of surplus water with the firm water costs, they achieve the dual purpose of recovering all costs while lowering the average cost to all users according to the amount of surplus water available.

A critical factor in surface water prices stems from the general concern over groundwater. Except in districts with no groundwater, a primary purpose of developing the surface water in the first place was to relieve pressure on groundwater supplies. It has always been recognized that pricing policies that resulted in higher prices for surface water than for groundwater would tend to defeat the purposes for which the surface water was initially obtained. Accordingly, it is a policy of the districts and of the Kern County Water Agency to keep the price of surface water at or below the price of groundwater.

The solution to this problem is particularly significant in Kern County, where the initial prices of state water when it was introduced into the county were higher than most prevailing groundwater pumping costs. Three principles govern the pricing policies in Kern districts with both groundwater and surface water supplies. First, the price of surface water to growers who use it exclusively is to be no greater than the costs of pumping groundwater in areas within the district where there is no surface water. Second, in areas with both surface and groundwater supplies, the price of surface water is established at levels lower than groundwater costs to induce the use of surface water. Third, in these joint supply areas the price of surface water is set close to groundwater pumping costs so as to encourage users to maintain wells and pumps for use during droughts and dry periods (Pyle, 1983).

In most instances, this strategy results in a surface water price that is not high enough to fully cover costs. The revenue shortfall is made up through additional property taxes on all users within the district. In this way, users who have access only to groundwater supplies help to pay the costs of surface water that is delivered elsewhere but that relieves the demand for water from a common aquifer. In return for the tax payment, groundwater users get lower pumping lifts and a consequent reduction in pumping costs. This pricing scheme results in effective conjunctive use of surface and groundwater supplies. It is by no means clear that the resulting array of prices is optimal in the economic sense, however, since no account is taken of the value of the marginal product of ground and surface water as they are currently employed.

The array of average prices for surface water in different districts is summarized in table 3-4, together with average groundwater pumping costs. These averages mask price differentials within districts because of variable pumping lifts and special pricing policies in joint supply areas. Average surface water prices reflect the annualized capital costs of district conveyance facilities, pumping costs, if any, and the wholesale water cost. This table illustrates the district-to-district variability in the proportion of cost assigned to the fixed and variable categories. In strict economic terms, this variability is completely arbitrary and distorts further the true costs of making water available to the user. However, a comparison of surface and groundwater costs does reveal that the latter are uniformly higher than the former, with the sole exception of the Cawelo Water District, where both surface and groundwater costs are among the highest found in the valley because of large pumping lifts.

The fact that the prices of surface water are structured to recover cost and induce conjunctive use of ground and surface water means that, in

TABLE 3-4. Average Retail Costs for Irrigation Water Per Acre-Foot in Selected Kern County Water Districts, 1981 (dollars)

District	Surface water			Average groundwater cost[a]
	Fixed costs	Variable costs	Total cost	
Arvin-Edison	25.00	12.00	37.00	78.80
Belridge	—	38.50	38.50	No Groundwater
Berrenda Mesa	29.60	56.12	85.72	No Groundwater
Buena Vista	3.50	7.25	10.75	46.05
Buttonwillow	19.90	37.00	56.90	57.31
Cawelo	54.25	38.00	97.25	92.81
Delano-Earlimart	4.24	7.13	11.37	38.23
Henry Miller	2.55	15.00	17.55	40.53
Kern Delta	4.00	10.21	14.21	52.89
Kern-Tulare	12.05	82.00	94.05	116.32
Lost Hills	15.90	19.78	35.68	No Groundwater
North Kern	1.95	20.00	21.95	62.72
Rosedale-Rio Bravo	No Surface Water			45.95
Semitropic	12.20	26.30	38.50	61.22
Shafter-Wasco	1.75	9.00	10.75	69.84
SSJMUD[b]	2.50	6.59	9.09	87.83
Wheeler Ridge	2.20	54.10	56.30	78.43

Source: Basic data were obtained from the individual irrigation districts.
[a] Includes annualized capital costs, standby charges, and variable cost.
[b] Southern San Joaquin Municipal Utilities District.

many instances, prices do not serve a rationing function. The basic allocation and rationing functions are accomplished through the formal system of entitlements (both legal and contractual) and ultimately through the availability of water. In years when water is plentiful, deliveries to districts and within districts are governed by contractual entitlements to both firm and surplus water and by the system of legal entitlements that governs the use of Kern River water. In water-short years, the rationing of surface water is usually governed by the availability of groundwater, with available surface supplies being directed to areas that have no access to groundwater.

In several districts, limitations on the capacity of conveyance facilities also serve to ration the quantity of surface water that can be delivered. This is most obvious in the Kern-Tulare Irrigation District and exists to a lesser extent in the Cawelo Water District. In addition, groundwater pumping in these districts is very expensive. Not surprisingly, high-valued crops predominate, underscoring the fact that groundwater costs are the principal economic rationing force in Kern County. The role of surface water pricing in promoting conjunctive use serves to dampen the rise of groundwater costs to some extent, although the trend of groundwater costs will inevitaby be upward as the price of energy rises. Without new supplies, groundwater costs will serve to increase the economic discipline imposed by water prices in the future.

Gains From Trade

It is clear that the extensive water conveyance and storage facilities in Kern County offer a significant potential for augmenting water supplies through trade. The physical exchange arrangements previously discussed illustrate this potential. These physical exchanges, which serve to alleviate local scarcities, appear to work only because certain districts have access to supplies that are surplus in the sense that cheaper water of the same class is available. To minimize costs, the water-surplus districts enter into exchange arrangements to avoid having to pay for all or some portion of their most expensive supplies.

But although these exchanges have economic attributes, real prices are not the driving force behind them. Rather, prices are strictly cost determined, and demand influences the direction and magnitude of water trade but not the price level. These arrangements differ somewhat from those in more conventional markets, where prices are established by the interaction of supply and demand and, as a result, more closely approximate true scarcity value.

The creation of markets in which water could be routinely bought and sold would provide an option for Kern County to adapt to the general scarcity created by groundwater overdraft, the ultimate inability of the state to deliver all the entitlement water, and the modest pressure for expansion of irrigated agriculture. The adaptation could occur in two distinct ways that are not necessarily exclusive. First, regional markets might allow water users within the county to import additional supplies from elsewhere in the state. Second, internal water markets would permit more rational economic exchange within the county. Such local exchanges would tend to ensure that water is put to its highest-valued uses, rather than being allocated predominantly by contracts and entitlements that are probably more the products of historical accident than reflections of the value of water in different uses at different locations. The potential gains to Kern County from market-like exchange in water have not previously been analyzed. There is, however, some corollary evidence that permits inferences to be drawn about such gains.

Interregional Trade

There are two pieces of evidence that provide insight into the potential gains to Kern County from trade that would permit its water users to acquire additional water supplies from other regions. The first relates to an attempt by the Kern County Water Agency, together with twelve districts within Kern County and several non-Kern agencies, to participate financially in the construction of Marysville Reservoir on the Yuba River in the northern Central Valley. The Kern County entities joined with the Yuba County Water Authority to complete engineering and financial studies of the feasibility of constructing Marysville Reservoir. The studies showed that the project would have an annual yield of 640,000 acre-feet, with 355,000 acre-feet of the total available for use in the San Joaquin Valley. The studies also showed that when account was taken of federal flood-control contributions and hydroelectric revenues, the cost to Kern County entities would be about $80 per acre-foot (Associated Engineering Consultants, 1983).

No final commitment was ever made to build this joint project because in 1981 Yuba County voters rejected a referendum on it by a vote of more than two to one. The reasons for the vote have not been carefully investigated. Apparently they were related to environmental concerns, worries about tax increases required to finance Yuba County's share, and general uncertainty about the long-term implications of reducing the water supply to Yuba County. It is not known how the relatively high cost of this water might have affected its attractiveness to Kern County because of Yuba's negative referendum. However, this episode illustrates that gains from trade with other regions are possible. Within Kern County, interest in such joint ventures has subsequently weakened because of political resistance in areas of origin and a tightening of federal cost-sharing requirements.

A second source of evidence is empirical. A study prepared for the water agencies of Kern County estimates that under current supply conditions (including prices), there is a net water deficiency of 370,000 acre-feet annually in Kern County, given current cropping patterns and irrigation practices (Associated Engineering Consultants, 1983). This total comprises 76,000 acre-feet of surface water deficiency in non-groundwater areas, 286,996 acre-feet of groundwater overdraft in agricultural areas, and 7,004 acre-feet of overdraft in the urban Bakersfield area. The surface water deficiencies represent average annual shortfalls in surface water deliveries to currently irrigated lands. The distribution of all the estimated deficiencies in agricultural water supplies is listed by district in table 3-5.

A conservative assessment of the willingness to pay for additional

TABLE 3-5. Estimated Annual Water Supply Deficiencies in Kern County Districts, 1980

Groundwater overdraft	Quantity (acre-feet)
Buttonwillow	12,517
Delano-Earlimart	2,680
Henry Miller	4,470
Kern Delta	75,103
North Kern	103,713
Pond Poso	9,835
Rosedale-Rio Bravo	4,470
Semitropic	23,246
Shafter-Wasco	13,411
SSJMUD[a]	10,729
Unincorporated Area	15,199
Wheeler Ridge	11,623
Subtotal	286,996
Surface water shortfall	
Belridge	5,400
Berrenda Mesa	6,400
Kern-Tulare	30,000
Lost Hills	30,900
West Kern	3,300
Subtotal	76,000
Total	362,996

Source: Adapted from Associated Engineering Consultants, *Report on Investigation of Optimization and Enhancement of the Water Supplies of Kern County* (Bakersfield, Calif.) January 1983.
[a] Southern San Joaquin Municipal Utilities District.

water in Kern County can be made by assuming that users in existing deficiency areas would be willing to pay at least as much for additional firm surface supplies as they pay for the last increment of water that they currently use. Since nearly 80 percent of this additional water would supplant currently mined groundwater, it is reasonable to assume that the value of the marginal product of this deficit water is at least equal to what growers pay for it. In groundwater mining areas, this would be equal to the marginal pumping cost. In nongroundwater areas, the case is less clear and the value of the marginal product is assumed to be equal to price paid for the last increment of existing surface water. This may overstate that demand, but probably not significantly, since nongroundwater areas account for only 20 percent of this shortfall. Based on these assumptions, the districts can be stratified by willingness to pay at their deficiency levels. An import-demand function, fitted to the resulting schedule of quantities and prices (dollars per acre-foot) has the following form, with quantities in millions of acre-feet and prices in 1980 dollars (F-statistics are noted in parentheses):

(1) $\quad\quad\quad Q = 0.592 \quad -0.006P \quad\quad R^2 = 0.96$
$\quad\quad\quad\quad\quad\quad (844.30) \quad (374.28)$

This import demand function can be used in a spatial equilibrium model developed by Vaux and Howitt (1984) for analyzing the potential of water markets. The model, which is of a type originally suggested by Enke (1951) for spatially separate markets, was previously employed to analyze trade among five aggregated regions within California. Specifically, the state was divided into three agricultural regions and two urban regions. Supply and demand functions were estimated for each region and subsequently used together with estimates of interregional transport prices to identify the direction, magnitudes, and benefits of trade among regions. The analysis was predicated on the assumptions that marginal cost-pricing practices would prevail and that groundwater could not be traded directly, although it could substitute for traded surface water supplies. (For a complete description of the model, see Vaux and Howitt, 1984.)

The model was extended in relatively straightforward fashion by disaggregating the southern agricultural region in which Kern County is located into a Kern County Region and a non-Kern County Region. The Kern County import-demand function was used, together with the supply-and-demand functions for the other regions, to assess the direction, magnitude, and gains from any trades that would occur. The results suggest that Kern County would be willing to buy 353,200 acre-feet from agricultural water users to the north at a price of $37.50 per acre-foot. The sellers would receive, in turn, $21.60 per acre-foot, with the $16 difference accounting for the costs of transport. This result is not surprising in view of the relatively high value of Kern County agriculture when compared with the value of agriculture in areas north of Kern County where water is more plentiful and cheaper.

The annual gains from trade are shown in table 3-6. For Kern County, the benefits have been computed separately for buyers in the groundwater and nongroundwater areas. The gains total nearly $10 million annually, based on 1980 conditions. Current estimates suggest that the existing deficiency could increase to 600,000 acre-feet by 1990 if the availability of surplus water delivered by the state is reduced. This suggests at least the possibility that the magnitude of gains from trade could grow over time. It is important to recognize that these gains accrue to current irrigation activity. The possibility that new water would be used to develop previously unirrigated lands has not been considered. How-

TABLE 3-6. Estimated Annual Gains From Interregional Trade Between Kern County and Northern Agricultural Region (dollars)

Region	Benefits
Northern agriculture	120,088
Kern County:	
Nongroundwater areas	1,244,654
Groundwater areas	8,357,294
Total	9,722,036

ever, there is no evidence to suggest that lands that are currently unirrigated could generate sufficient productivity to bid water away from users north of Kern County.

The existence of substantial gains from trade between Kern County and the Central Valley regions to the north is not surprising in view of the relative preponderance of high-valued perennial crops, vegetables, and cotton in Kern County. This analysis shows that trading in interregional markets could alleviate all but about 10,000 acre-feet of Kern County's existing 370,000 acre-foot water deficiency. Kern County might adjust to this remaining deficiency, or others occasioned by barriers to interregional trade, through an internal market in which water could be traded within the county.

Internal Trade

The physical exchanges of water already described attest to the capacity of Kern County's plumbing system to accommodate internal movements of water. In general, there is both an elevation gradient and a "price gradient" flowing from east to west in the valley portion of the county. The elevation gradient is significant, since exchanges along it can be accomplished with minimal transport costs, whereas exchanges against it require the purchase of energy for pumping. The Associated Engineering Consultants (1983) have identified at least two further exchanges. In these, the Kern Delta and North Kern districts upstream could receive some of the Kern River entitlements of the Henry Miller and Buena Vista districts in exchange for entitlements to water from the State Water Project whose aqueduct abuts the latter two districts. These exchanges would bring a little more water to Kern County and would minimize the pumping required to deliver surface water to all four of these districts. The exchanges have not occurred because North Kern has no entitlement to state water and there is uncertainty over the ability of the state to meet Kern Delta's entitlement in future years.

TABLE 3-7. Surface Water Prices of West Side and East Side Water Districts in Kern County, 1981 (dollars)

West Side		East Side	
District	Price	District	Price
Buena Vista	10.50	SSJMUD[a]	9.09
Henry Miller	17.55	Shafter-Wasco	10.75
Lost Hills	35.68	Delano-Earlimart	11.37
Belridge	38.50	Kern Delta	14.21
Semitropic	38.50	North Kern	21.95
Wheeler Ridge	56.10	Arvin-Edison	37.00
		Kern-Tulare	94.05
		Cawelo	97.25

[a] Southern San Joaquin Municipal Utilities District.

The elevation gradient that makes physical exchanges attractive is reinforced by a price gradient, which suggests that gains from trade through market-like arrangements may also be attractive. The price gradient is illustrated in table 3-7, where west side and east side districts are listed in order of increasing water costs. If the Kern-Tulare and Cawelo districts are ignored and it is recognized that there is generally a north-to-south trend in prices within each group, it becomes clear that water prices on the east side of the valley are substantially lower than those on the west side. Moreover, the fact that water can be conveyed by gravity through the Friant-Kern Canal and the Kern River means that transport costs are likely to be quite low, perhaps on the order of $5.00 per acre-foot.

Although no analysis of potential gains from market-like trade in Kern County exists, it seems clear from the available price differentials and the potentially low transport costs that such trade could occur. However, the extent of this trade would be tempered, maybe severely, by certain institutional factors. Price differentials are influenced significantly by the differences in the wholesale price of water charged by the Bureau of Reclamation and the state of California, respectively. The Bureau of Reclamation, which is the primary supplier on the east side of the valley, charges $3.50 per acre-foot for firm water; the average 1981 wholesale price of water to the Kern County Water Agency, which primarily services west side districts, was $21.12 per acre-foot (Kern County Water Agency, 1982). The fact that the Bureau of Reclamation subsidizes water by forgiving interest payments and using revenues from hydroelectric power generation to underwrite project costs, is well known. The artificially low cost of water to districts served by the Friant-Kern Canal on the east side could be reflected in the production of a low-valued crop mix when compared with crop-mix values produced in districts with higher water costs.

An analysis of crop values in the various Kern County districts reveals differences that only partially reflect differences in the price of water (Vaux, 1983). This suggests the existence of economic rents for water, rents that have been capitalized into the value of land. No evidence has been gathered to support or refute the water-rent hypothesis, but the existing data are strongly suggestive of their existence (Gardner, 1983). If east side growers receive water rents, the price differentials shown in table 3-7 might overstate the extent of potential gains from trade, since prevailing prices may be less than the value of the marginal product of the water.

The conclusion with respect to the potential of internal markets in Kern County is not clear-cut. The presence of both an elevation and a water price gradient running from east to west strongly suggests the existence of trade gains from an internal market. While such gains almost assuredly exist, they may be tempered by capitalization of low water prices into land values and the consequent loss in those values. Ultimately, however, trading would certainly occur at the margin, and Kern County could improve the efficiency with which its water is allocated internally by establishing a local market. This would help reduce economic losses from any future adjustment to increasing groundwater costs and anticipated shortfalls in surface water deliveries.

Barriers to Trade

The various water agencies in Kern County have pursued the possibilities for water trading more aggressively than agencies in other parts of the state, as indicated by the existing exchange arrangements. Yet the evidence in the foregoing section suggests that incentives for further trade both with other regions and internally still exist in the form of price differentials. The lack of response to these incentives is attributable to various barriers to trade in the law and other institutions. Phelps and coauthors (1978) conducted a survey of water trades that did occur during the 1976/77 drought in California and found that virtually all such trades had two characteristics: first, the selling party had clear title to the water being traded, and second, the trades were not subject to review and approval by the State Water Resources Control Board. In general, then, the principal barriers to trade are imperfections in the California

Water Rights Law and the potentially high transactions costs associated with water trades. These barriers take a variety of forms.

Water Rights

Water held by private individuals under the riparian and appropriative doctrines is likely to be especially difficult to trade because of legislative restrictions and, in some instances, uncertainty of the extent of the right. As noted earlier, riparian rights attach to riparian lands. Any transfer of the water from these lands ends the right. Thus, riparian rights and their counterpart correlative rights to groundwater are not transferable to new locations. Appropriative rights, on the other hand, are not necessarily appurtenant to the land, but they cannot be transferred if injury will result to downstream users. Phelps and coauthors (1978) concluded that the mere presence of third-party users might be enough to block a sale of appropriative water.

A crucial barrier to the trade of water held under these rights relates to the uncertain quantity of water to which the right holder is entitled. Riparian rights are subject to "substantial variation in quantity" (Lee, 1977), whereas the magnitude of pre-1914 appropriative rights has not been recorded. The magnitude of appropriative rights established between 1914 and 1969 is similarly uncertain, although flow rates and seasonal restrictions are recorded. These uncertainties cloud the nature of private ownership sufficiently to inhibit exchange. An owner seeking to sell an appropriative right invites an administrative or legal review that may result in diminution of the right but cannot increase it, a risk that discourages such sales.

The stipulation that appropriative and riparian water be put to "reasonable and beneficial use" is also troublesome, because the concept of reasonable and beneficial is essentially undefined (Phelps and coauthors, 1978). This creates the possibility that the act of selling water may be taken as evidence of non-use and lead to forfeiture. Historically, the stipulation has served to restrain efforts to use water efficiently at the field level by eliminating any returns to the right holder other than his cost savings, which are often minor. Recently adopted legislation (Assembly Bill 3491, 1982) stipulates that rights cannot be lost when a water-right holder reduces use and sells water. Such proposed sales must be submitted to the Water Resources Control Board for approval, which must be based on a finding that the proposed transfer would have no adverse effects on the economy of the area from which the water is being transferred. It is too early to assess the impact of this legislation, although it appears to have removed forfeiture penalties as a barrier to trade. It may not have significantly reduced the transactions costs arising from administrative review of exchanges, however.

Third-Party Effects and Transactions Costs

The interrelatedness of water use means that third parties must be protected in any trade. To ensure this, trades must be approved by the Water Resources Control Board. The board's approval procedures are time consuming, and when challenges are involved, can be costly. Moreover, as noted, there may be a presumption that trades will be disallowed simply because of the existence of third-party users. Under these circumstances, a potential seller is likely to decide that the transactions costs of any trade can be very high relative to the probability of obtaining the approvals necessary to consummate the trade.

A common recommendation for resolving the technological externalities of water trading is to limit the quantities of water that can be traded to the amount that is used consumptively. A difficulty here is that levels of consumptive use are not known and may be subject to dispute. Letey and Vaux (1984), for example, calculated the annual consumptive use for six irrigation districts in California and found that the calculated consumptive use in an average year exceeded the average annual water applied in the districts in question. The quantification of water rights and consumptive use would no doubt require litigation and the costs of accomplishing this on a broad scale, such as for Kern County, would be high. The director of the Department of Water Resources has cited an estimate of $50–$100 million simply to acquire the data needed to adjudicate groundwater rights in the San Joaquin Valley (Kennedy, 1983).

The uncertainties surrounding the quantification of most private water rights, and the need to protect third parties in any trade, represent barriers to trade that may be both costly and difficult to eliminate. On balance, these barriers are substantial enough to suggest that general trading in private water rights may be limited. Transactions costs may not serve to inhibit trade between water districts, however.

Trade Constraints on Water Districts

The various types of water districts that retail or wholesale water could become the focus of a water market with trades occurring between districts. There are two ways in which this might occur. Phelps and coauthors (1978) suggest that districts should convey clear title to each of their member users. The individual users would then be free to buy and sell water as need arose. Alternatively, the districts themselves could act as water brokers, receiving and posting offers and brokering trades between member users and external entities. There are at least two major barriers, to this kind of trade, however.

The first is a legal restriction on outside sales by irrigation districts. In general, the law holds that districts exist primarily to supply water only to members and that only water that is "surplus" may be sold outside the district. Further, a contract for the sale of surplus water cannot extend for more than seven years (California Water Code, Secs. 22250–22257, 22259, and 22109 as amended). This provision is evidently intended to protect districts from any financial uncertainties that might arise from trade and ensure that they can meet their contractual obligations and commitments for bonded indebtedness (Howitt and coauthors, 1982). Moreover, it has been pointed out that external sales of water by districts may foster the assumption of a public-service responsibility to continue supplying water to external buyers indefinitely and may jeopardize the stability of internal water supplies (Bain and coauthors, 1966).

A second major barrier stems from the legislative restraint on profits by irrigation districts. If water is sold at marginal cost, as it will probably have to be to provide sufficient incentive, districts will be put in a profit-making position. Possible remedies for this have not been well explored, but they include changes in legislation and possibly a system of rebating profits to water users, either directly or in the form of tax relief. Indeed, strictures against profit making appear to run counter to the need to protect financial integrity. A system of profit rebates to individual water sellers would meet the spirit of the law, and it is difficult to think of parties who would have an incentive to challenge such a system legally.

Although these barriers to water markets have not been subjected to exhaustive legal analysis, it appears that most issues could be resolved under existing legislation. The notion of the irrigation district as a public-service entity and the obligations that flow from that concept might appear especially troubling. Yet recent legislation (Assembly Bill 3491, 1982) makes clear the legislature's support for water trading and, by implication, denies the possibility that a selling party could incur some public service obligation in perpetuity. The seven-year restriction on any sale would also appear to deny that public-service obligations may attach to trade.

The seven-year limitation on water sales represents a barrier in itself, however. The discussion of existing exchanges in Kern County illustrates that firm water supplies are often sought. The seven-year limitation compromises the firmness of purchased supplies and limits the potential of water markets to respond to demands for firm supplies in the long run. It seems clear that removal of this barrier will require legislation.

Although current legal institutions do not appear to inhibit market-like water exchanges except in a long-term sense, they clearly do not foster it. The evolution of western water law placed heavy emphasis on security of tenure in order to promote settlement of the land. This helped create attitudes about water use that are antithetical to market exchange. The fact that market-like exchanges of water have not been challenged legally creates uncertainty for districts proposing to buy or sell water on anything but an immediate, short-term basis. Current laws will foster trade only after the development of a body of case law that defines clearly the circumstances under which exchanges in water can occur. The absence of wide experience with market exchanges of water under a variety of circumstances appears to constrain such exchanges more than anything else.

Conclusions

The water scarcity problem in Kern County is imposed by inadequate supplies of cheap irrigation water. The state's inability to deliver the full amount of its contractual water obligations is viewed by Kern County water users as the major cause of this scarcity. Groundwater overdraft is the main symptom. There appear to be at least three ways in which the situation could resolve itself or be resolved. First, it is possible that neither new supplies nor the institutional changes necessary to permit water marketing will be realized. Under this no-change scenario, groundwater overdraft is likely to continue and to increase if water now surplus to the needs of the Los Angeles basin becomes unavailable. Similarly, economic losses associated with the elimination of groundwater overdraft would likely grow from the current estimated value of $24 million annually, and some economic dislocations would result.

A second resolution would involve the construction and operation of additional water supply facilities, probably by the State of California. The increasing costs of such facilities, coupled with the competition for scarce water supplies from other sectors and political resistance to further public underwriting of project costs, make this option unlikely. In addition, current estimates of the costs of these new supplies suggest that they will be in excess of $100 per acre-foot and raise questions about the ability and willingness of agricultural users to defray them fully. The mounting economic and political costs of developing new supplies suggest that this may be the most expensive way to deal with Kern County's water problem. Moreover, if new supplies were used to bring new land into production rather than as a substitute for groundwater overdraft, the water scarcity situation would remain unchanged.

A final resolution involves making the legal and institutional changes necessary to create water markets. Evidence presented in this chapter suggests that Kern County could trade profitably with other regions and thereby acquire a substantial portion of the water necessary to make up the projected shortfall. This trading potential is evidence that the marginal economic contribution of Kern County agriculture exceeds that of some irrigated lands to the north. Thus, the losses to Kern County agriculture in the event that new supplies do not become available will likely be greater than those accruing elsewhere as a consequence of a water shortfall. The development of water markets, however, would eliminate the need for some group or sector to incur uncompensated economic losses as a consequence of water scarcity. In short, a market system could aid in resolving Kern County's water problem by removing existing inefficiencies in water use, to the benefit of both water buyers and sellers.

References

Arvin-Edison Water Storage District. 1983. *The Arvin-Edison Water Storage District Water Resources Management Program* (Arvin, Calif., August).

Associated Engineering Consultants. 1983. *Report on Investigation of Optimization and Enhancement of the Water Supplies of Kern County* (Bakersfield, Calif., January).

Bain, Joe S., Richard E. Caves, and Julius Margolis. 1966. *Northern California's Water Industry* (Baltimore, The Johns Hopkins Press).

Bowden, Gerald D., Stahrl W. Edmunds, and Norris C. Hundley. 1982. "Institutions: Customs, Laws and Organizations," in Ernest A. Engelbert and Ann Foley Scheuring, eds., *Competition for California Water* (Berkeley, Calif., University of California Press).

California Department of Water Resources. 1978. *General Comparison of Water District Acts*, Bulletin 155-77 (Sacramento, Calif., DWR)

Enke, S. 1951. "Equilibrium Among Spatially Separated Markets: Solution by Electric Analogue," *Econometrica*, vol. 19, no. 1, pp. 40-47.

Gardner, B. Delworth. 1983. "Water Pricing and Rent Seeking in California Agriculture," in Terry L. Anderson, ed., *Water Rights* (Cambridge, Mass., Ballinger Publishing Co.).

Governor's Commission to Review California Water Rights Law. 1978. *Final Report* (Sacramento, Calif.).

Hirshleifer, Jack, James DeHaven, and Jerome Milliman. 1960. *Water Supply: Economics, Technology, and Policy* (Chicago, University of Chicago Press).

Howitt, Richard E., Dean E. Mann, and H. J. Vaux, Jr. 1982. "The Economics of Water Allocation," in Ernest A. Engelbert and Ann Foley Scheuring, eds., *Competition for California Water* (Berkeley, Calif., University of California Press).

Hutchins, Wells A. 1956. *The California Law of Water Rights* (Sacramento, Calif., Office of the California State Engineer).

Kennedy, David N. 1983. "The Role of Groundwater in California Water Policy," *Proceedings of the Fourteenth Biennial Conference on Ground Water* (Davis, Calif., California Water Resources Center).

Kern County Agricultural Commissioner. 1981. *Agricultural Crop Report, Kern County* (Bakersfield, Calif.).

Kern County Water Agency. 1982. *Biennial Report 1980–81* (Bakersfield, Calif., February).

———. 1983. *Water Supply Report, 1982* (Bakersfield, Calif., May).

Lee, Clifford T. 1977. "The Transfer of Water Rights in California: Background and Issues," Governor's Commission to Review California Water Rights Law, Staff Paper No. 5 (Sacramento, Calif., State Printing Office).

Letey, J., and H. J. Vaux, Jr. 1984. "Water Duties for California Agriculture." Report prepared for California Water Resources Control Board (Sacramento, Calif., July).

Northwest Economic Associates. 1983. *The Impact of the State Water Project in Kern County: 1983 Economic Study* (Vancouver, Wash., May).

Phelps, Charles E., Nancy Y. Moore, and Morlie H. Graubard. 1978. *Efficient Water Use in California: Water Rights, Water Districts, and Water Transfers* (Santa Monica, Calif., The Rand Corporation).

Pyle, Stuart T. 1983. "Economic Strategies for Kern County Groundwater Management," *Proceedings of the Fourteenth Biennial Conference on Ground Water* (Davis, Calif., California Water Resources Center).

State of California. 1980. *California Statistical Abstract 1980* (Sacramento, Calif., California Documents Section).

Vaux, H. J., Jr. 1983. Unpublished data.

———, and Richard E. Howitt. 1984. "Managing Water Scarcity: An Evaluation of Interregional Transfers," *Water Resources Research*, vol. 20, no. 7, pp 785–792.

Watson, William D., Carole Frank Nuckton, and Richard E. Howitt. 1980. *Crop Production and Water Supply Characteristics of Kern County*, Bulletin 1895 (Davis, Calif., Giannini Foundation, April).

4
Satisfying Southern California's Thirst for Water: Efficient Alternatives

*Richard W. Wahl and Robert K. Davis**

The southwestern corner of the United States is one of the driest parts of the country. Although rainfall averages about 3 inches per year, some 450,000 acres are irrigated in the extremely productive Imperial Irrigation District and another 100,000 acres in the Coachella Valley County Water District farther to the north (see figure 4-1). Between the Coachella and Imperial districts lies the Salton Sea, a saline inland lake that drains both the Imperial and Coachella valleys.

Climatic and hydrologic conditions to the west on the southern California coastal plain, where Los Angeles and its suburbs are located, are only somewhat more generous. Rainfall averages less than 20 inches per year, and all the coastal rivers are seasonal, with no flow in the summer months. Yet it is here that one of the major metropolitan centers of the country has grown and prospered. The Metropolitan Water District of Southern California (MWD) now provides water to twenty-seven member agencies with a combined service area of 5,100 square miles.

This growth would have been impossible had it not been for the vision and daring of a few notable early twentieth-century water engineers. Los Angeles and MWD are now able to draw water through three very

---

*Respectively, economist and assistant director for economics, Office of Policy Analysis, U.S. Department of the Interior. Robert Davis spent nine months during 1983–84 studying California water resource problems under the sabbatical program of the Federal Senior Executive Service. The views expressed in this paper do not necessarily coincide with any official positions of the Department of the Interior.

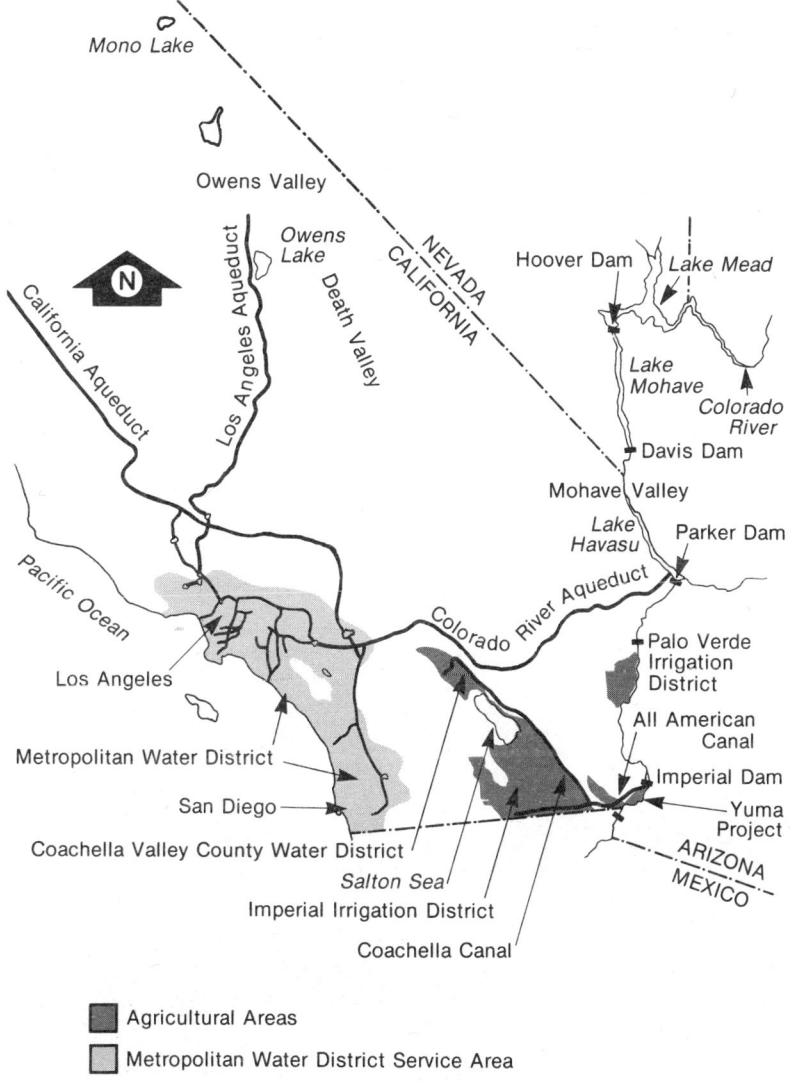

Figure 4-1. California developments on the Colorado River *Source:* Colorado River Board of California, *Annual Report, 1963–64* (Los Angeles, Calif., 1964).

long straws: from the northeast, the 240-mile Los Angeles Aqueduct draws water from the Owens Valley and from the Mono Lake Basin; from the east, the 242-mile Colorado River Aqueduct carries water from Lake Havasu behind Parker Dam; and from the north, the 444-mile California Aqueduct pumps water from the Sacramento-San Joaquin

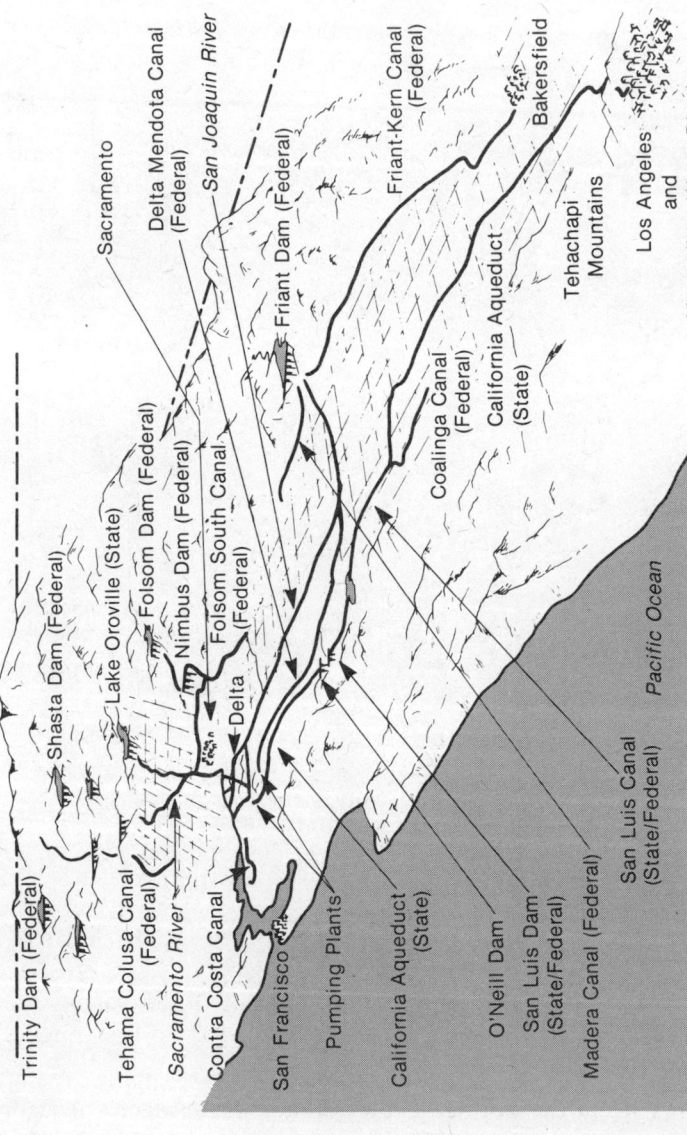

Figure 4-2. Central Valley Project and State Water Project *Source*: Adapted from map of the Central Valley Project, U.S. Department of the Interior, Bureau of Reclamation

Delta (figure 4-2). Such long-distance transport has not been accomplished without political opposition. Los Angeles has been unable to live down its reputation as a robber baron of the water rights and agricultural production of the Owens Valley. The city is now faced with the loss of a significant amount of Mono Lake water and may face restrictions on further groundwater pumping in Owens Valley. The MWD has relied heavily on expected additions to the State Water Project (SWP), but on June 9, 1982, California voters rejected major additional transport of water from north to south in the state by rejecting Proposition 9, which would have permitted construction of the Peripheral Canal across the delta. In addition, although the MWD has been able to use surplus flows from the Colorado River to meet some of its demands, that supply will be substantially reduced because in 1985 Arizona began to make increased use of its share of water under the Colorado River Compact. The MWD is projecting that its supplies will fall short of demand before the year 2000.

In the early twentieth century, the city fathers of Los Angeles were able to secure additional water supply, but, in the face of widespread public interest in water supply and other environmentally related matters, current options for MWD are limited. After reviewing existing water sources, this chapter will examine options available to MWD for balancing demand and supply: (1) additions to the SWP, (2) purchase of surplus water from the federal Central Valley Project (CVP), (3) demand management, (4) water conserved by investment in the Imperial Irrigation District, and (5) other alternatives.

The objective is to compare the costs of the four principal alternatives mentioned and to examine the prospects for successful implementation of the more economically efficient alternatives. However, to comprehend the institutional forces that impinge upon the success of these alternatives, it is necessary to understand the deep-rooted attitudes deriving from three-quarters of a century of conflict over water supplies in California—a conflict that has often centered on Los Angeles water engineering and real estate practices. It is from this perspective that we first review the availability of water from local supplies, as well as supplies from the three aqueducts now serving the Los Angeles area.

Existing Sources of Supply

Local Supplies

Local supplies were, of course, the first to be developed by cities on the southern California coast. Surface supplies are quite limited, and widespread groundwater pumping led to concerns about overdrafting as early as the 1920s. Court adjudication of groundwater basins currently limits the amount of groundwater extraction in most areas to the amount replenished each year. The natural recharging of groundwater is supplemented by artificial recharge using reclaimed water and water imported by MWD.

Table 4-1 shows the volume of water expected from locally developed sources in the six counties supplied by MWD. As the table indicates, MWD estimates supply for the year 2000 under three hydrologic conditions: (1) average-year supply, (2) supply available over a prolonged dry period—the 1928 to 1934 hydrologic conditions, and (3) supply available during a severe short-term drought, such as during the 1976/77 period. Local sources are expected to deliver approximately the same supply levels in the projection period under all hydrologic conditions. This reflects the fact that these sources are nearly fully developed and that they rely heavily on groundwater pumping and so are expected to be little affected by variations in rainfall (about 0.1 maf comes from surface sources and 1.0 maf from groundwater). The local supply entries in the table exclude water imported by MWD for groundwater replenishment, since these amounts are included in other entries in the table.

TABLE 4-1. Metropolitan Water District Estimates of Service Area Supply and Demand
(million acre-feet per year)

	Average-year supply		Repeat of 1928–1934 dry period		Repeat of 1976–1977 drought	
	1985	2000	1985	2000	1985	2000
Supply						
Local	1.11	1.12	1.11	1.12	1.11	1.12
Los Angeles Aqueduct	0.47	0.47[a]	0.47	0.47[a]	0.30	0.30[a]
Colorado River	0.46	0.40	0.46	0.40	0.46	0.40
State Water Project	1.51	1.48	1.13	0.99	0.71	0.59
Total	3.55	3.47	3.17	2.98	2.58	2.41
Demand						
"Normal demand"	3.19	3.61	3.19	3.61	3.19	3.61
MWD's estimate of deficit	—	0.14	0.02	0.63	0.61	1.20

Sources: Metropolitan Water District of Southern California, *Water Supply Available to Metropolitan Water District Prior to Year 2000*, Report No. 948 (Los Angeles, Calif., 1983), p. I-4, table 1, and *1982 Population and Water Demand Study*, Report No. 946 (Los Angeles, Calif., 1982) p. IV-3, table 4 and p. VIII-4, table 16.

[a] Subject to possible decrease of 0.1 to 0.2 maf because of Mono Lake and Owens Valley decisions.

Owens Valley and the Los Angeles Aqueduct

William Mulholland, superintendent of the Los Angeles water department from 1902 to 1928, had a clear vision of the limits of local water supplies and devised bold measures to secure additional water. On July 29, 1905, the Los Angeles *Examiner* announced that the city had obtained the necessary options to bring a new water supply, "immense and unfailing," some 240 miles from the Owens River.

The residents of Owens Valley had entertained other expectations for their water. (Historical accounts of the conflict between Owens Valley and Los Angeles over water supply are available in Nadeau, 1950; Hoffman, 1981; and Kahrl, 1982.) As early as 1903, Reclamation hydrologists were surveying the Owens Valley as the site for an irrigation project. Hopes ran high for a project that would ensure the area's agricultural development, and Frederick H. Newell, chief engineer of the Reclamation Service, received numerous letters from valley residents enthusiastic about the project. Newell's polite responses to these letters did not reveal that he had already been informed by J. B. Lippincott, the Reclamation Service chief in California, that Los Angeles was also interested in the Owens Valley as a source for water. Lippincott, a resident of Los Angeles and a friend to Mulholland, agreed to keep the city's intentions unknown to Owens Valley residents (Hoffman, 1981).

When Los Angeles' water importation plans were announced, Owens Valley residents were understandably outraged. To secure the riparian rights, Fred Eaton, a former Los Angeles mayor, had purchased land options while posing as a purchaser for a cattle rancher. Furthermore, many of the local residents assumed that Eaton was actually securing water rights not for Los Angeles but for the Reclamation Service project. When Lippincott's role in assisting Los Angeles became known, he was denounced by Owens Valley residents as an accomplice. In 1906, Lippincott resigned from the Service to become Mulholland's assistant, and this action only fueled impressions of Lippincott's impropriety.

In the period from 1905 to 1910, the city purchased some 70,000 acres of Owens Valley properties, and in 1913 the Los Angeles Aqueduct was placed in operation. These early purchases were in the southern part of the valley, where there was very little agricultural development. However, when the latter part of the 1910s brought a cycle of dry years, the city turned to purchasing land in the northern part of the Owens Valley, an area of existing agricultural development. With these purchases it is said that the real water wars in Owens Valley began (Nadeau, 1950). When Los Angeles purchased the McNally ditch properties in 1923, Owens Valley farmers began diverting the water before it entered the ditch. Los Angeles retaliated by intensifying its purchases of valley property. There were allegations that the city bought land in a checkerboard pattern and then, by failing to maintain irrigation ditches, forced other farmers into ruin. It was during the mid-1920s that the Los Angeles Aqueduct was dynamited several times, and shotgun guards, patrolling with searchlights at night, were used to keep it in operation. With the decrease in irrigation water supplies, the economy of the Owens Valley gradually shifted from dependence upon agricultural production to dependence upon recreation, such as hunting and fishing.

President Theodore Roosevelt commented on the Owens Valley controversy by saying that a broad view of the nation's resource development required that the valley be sacrificed to the "infinitely greater interest" of the city (Hoffman, 1981). A benefit-cost calculation would probably indicate Roosevelt's conclusion to be correct. Although Owens Valley possesses an abundant water supply, some of the soils are alkaline, the growing season is shorter than in lower areas, and the region was far from major population centers and without rail transport. However, Los Angeles could have avoided much of the conflict with Owens Valley farmers by developing a water supply project that would have enhanced agricultural production in the valley, as well as providing water for Los Angeles. Such a project was indeed considered and would have involved construction of a storage reservoir at the Long Valley site in the northern part of Owens Valley. Mulholland refused to press for this alternative when the owner of the large Long Valley ranch, Fred Eaton, demanded an exorbitant price of $1 million.

Litigation initiated since 1973 could reduce water supplies obtainable from the Los Angeles Aqueduct. Groundwater pumping in the Owens Valley was increased after completion of a second Los Angeles aqueduct in 1970. In 1973 the Inyo County Board of Supervisors filed a lawsuit against Los Angeles in an attempt to restrict the pumping. The board was concerned that the water table would be lowered below the root zone of the native vegetation, thereby causing a dramatic decline in the natural beauty of the area. The city of Los Angeles and Inyo County have reached a settlement under which a five-year study of local groundwater effects will be conducted in conjunction with the U.S. Geological Survey.

The Mono Basin extension of the Los Angeles Aqueduct, which was placed in operation in 1941, diverts water from four of the five major freshwater streams flowing into Mono Lake. Between 1941 and 1971 the level of Mono Lake dropped about one foot a year. After Los Angeles placed its second aqueduct in operation, the water level began decreasing by 1.6 to 2 feet per year. The lake is a nesting site for 90 percent of the state's population of California gulls, which cross the Sierras to nurture their young on two large islands in the lake. The decrease in freshwater inflow will eventually produce salinity levels that will adversely affect the brine shrimp on which the gulls feed. The subsiding water level has already turned Negit Island into a peninsula, and the island has been abandoned as a nesting site because nests there are now subject to attacks by coyotes.

An intergovernmental task force recommended excavation of a channel to prevent predators from reaching Negit Island and an immediate reduction of water exports to Los Angeles from the current 100,000 acre-feet to 15,000 acre-feet, among other measures. In 1980 several bills were introduced into the state legislature mandating implementation of the task force recommendations, although none was passed. The National Audubon Society, the Los Angeles Audubon Society, Friends of the Earth, and the Mono Lake Committee filed suit against Los Angeles to halt diversions from the lake until a suitable lake level is determined. The case has undergone a number of appeals, which have not yet been exhausted.

Table 4-1 shows a stable supply of 470,000 acre-feet from the Los Angeles Aqueduct, except during a repeat of the 1976/77 drought conditions under which this supply drops to 300,000 acre-feet. Under normal conditions, about 100,000 acre-feet a year come from the Mono Basin and about 200,000 acre-feet come from Owens Valley. Although the supply is shown as remaining at historical levels in the MWD projections, the Mono Lake litigation could conceivably reduce those supplies by as much as 85,000 acre-feet, and the Owens Valley settlement could eventually lead to reductions in groundwater pumping. The MWD believes that reductions from these two sources could total as much as 100,000 to 200,000 acre-feet (MWD, 1983b).

The Colorado River and the Colorado River Aqueduct

Drought in the early 1920s led Los Angeles to look not only for immediately developable water supplies in the northern part of Owens Valley but also for eventual development of supplies from the Colorado River. Arizona and the states on the upper Colorado made clear that they would oppose construction of a storage reservoir on the Colorado that could support diversions to California until the states bordering on the river reached an accord apportioning its waters. State representatives reached agreement on the Colorado River Compact in 1922, providing for the "equitable division and apportionment" of the river water based on "beneficial consumptive use" of 7.5 maf per year to the Upper Basin states of Wyoming, Colorado, Utah, and New Mexico and 7.5 maf to the Lower Basin states, California, Arizona, and Nevada. By 1928 the California forces in Congress were successful in gaining passage of the

Boulder Canyon Act, which authorized the construction of Hoover Dam and the construction of an irrigation canal from the Colorado River to the Imperial and Coachella valleys. The act also allocated Colorado River water among the Lower Basin states in the following shares: California, 4.4 maf; Arizona, 2.8 maf; and Nevada, 0.3 maf per year. In addition, California and Arizona are each entitled to 50 percent of any surplus flows.

The 1931 "Seven-Party Water Agreement" established priorities within California's share of Colorado River water (see table 4-2). As the table shows, agricultural uses were generally given priority over municipal uses. The cumulative totals in table 4-2 show that California's 4.4 maf share is exhausted after the fourth priority of 0.55 maf for the Metropolitan Water District. As table 4-3 indicates, California water users have been using more than their 4.4 maf share for a number of years. Of all the California users, MWD has taken the most advantage of surplus flows to California. For example, during 1977 MWD used 1.277 maf, reflecting reliance upon Colorado River water instead of water originating in the northern part of the state during the 1976/77 drought. Use of flows in excess of 4.4 maf by California water contractors has

TABLE 4-2. Priorities for Colorado River Water in California (Seven Party Agreement)

Priority	Water-using entity	Allocation (maf)	Cumulative total (maf)
1.	Palo Verde Irrigation District (For use on 104,500 acres)		
2.	Yuma Project (For use on 25,000 acres)		
3.a.	Imperial Irrigation District (IID) & Coachella Valley County Water District (CVCWD)	3.85	3.85
b.	Palo Verde Irrigation District (For use on an additional 16,000 acres)		
4.	Metropolitan Water District	0.55	4.40
5.a.	Metropolitan Water District	0.55	4.95
b.	City and County of San Diego	0.112	5.062
6.a.	IID and CVCWD		
b.	Palo Verde Irrigation District (For use on an additional 16,000 acres)	0.3	5.362

Source: U.S. Department of the Interior, Bureau of Reclamation, *Hoover Dam Power and Water Contracts and Related Data* (Washington, D.C., 1950) pp. 283–287.

TABLE 4-3. Use of Colorado River Water[a]

	Annual average 1971–1981 (1,000s acre-feet)
Lower Basin	
California	
Palo Verde Irrigation District	447
Yuma Project	62
Imperial Irrigation District	2,930
Coachella Valley County Water District	517
Total agricultural[b]	3,956
Metropolitan Water District	975
Indian reservations[c]	17
Miscellaneous[d]	32
Total California	4,980
Arizona	1,267
Nevada	79
Total Lower Basin	6,326
Upper Basin depletions	3,618
Total U.S.	9,944
Deliveries to Mexico	2,382
Total[e]	12,326

Sources: Lower Basin net diversions are from annual reports titled "Compilation of Records in Accordance with Article V of the Decree of Supreme Court of the United States in *Arizona* vs. *California* dated March 9, 1964," U.S. Bureau of Reclamation, Boulder City, Nevada. Upper Basin depletions and deliveries to Mexico are from *Annual Reports* of the Colorado River Board of California (Los Angeles, Calif.) as reported in Environmental Defense Fund, *Trading Conservation Investments for Water* (Berkeley, Calif., Environmental Defense Fund, Inc., 1983) p. 12, table 3.
[a] Values given are net diversions (diversions less measured returns).
[b] First three priorities under Seven Party Agreement.
[c] Fort Mojave and Colorado River Indian Reservations.
[d] Needles, Blythe, and others.
[e] Totals may not add due to rounding.

been possible because other Colorado River water users were not prepared to use their full shares.

However, the Supreme Court's 1963 opinion in *Arizona* v. *California* and its 1964 decree affirmed the allocations of the Boulder Canyon Act, and in 1968 the means for Arizona to use its allocation was provided by the Colorado River Basin Project Act. This act authorized construction of the Central Arizona Project (CAP) to divert Colorado River waters to the Phoenix and Tucson areas. The CAP has been under construction since 1970 and began diverting water in late 1985. Although more time may elapse before CAP diversions increase to their full entitlement, such an increase will mean that MWD's share will drop to 550,000 acre-feet, some 425,000 acre-feet below the annual average use for 1971 to 1981.

During passage of the Colorado River Basin Project Act, California won an important priority for its allocation over that of Arizona. The Colorado River Compact provides for up to 15 maf of use by the Upper and Lower basins, and the Mexican Water Treaty of 1944 guarantees an additional 1.5 maf to Mexico. However, annual flows of the Colorado have averaged less than 14 maf. To date, this overallocation has had little impact because Upper Basin use is far short of 7.5 maf (refer to table 4-3), but these figures indicate that the full compact and treaty allocations cannot be satisfied on a long-term basis. However, the abundant storage capacity on the Colorado River and California's priority virtually ensure that California water users, including MWD, will not have to reduce their allocations to contribute toward meeting Mexico's allocation in low-water years. For these reasons the supplies from the Colorado River system shown in table 4-1 are invariant with respect to California hydrologic conditions. MWD's 550,000 acre-foot Colorado River allocation is reduced by Indian water claims, conveyance losses, and other miscellaneous uses, resulting in net deliveries of 460,000 acre-feet. By the year 2000, MWD projections indicate an increase in Indian water use and in use of water by thermal power plants in the desert which would further reduce MWD's annual deliveries from the Colorado River to 400,000 acre-feet.

The State Water Project and the California Aqueduct

In the Central Valley of California the dry period from 1928 to 1934 led to increased pumping of groundwater, but the high cost of sinking deeper wells and the increased costs of pumping sent many farmers into bankruptcy. There was an appeal to the state to provide a water supply plan for the valley (Kahrl, 1978). An act approving the state Central Valley Project was approved on December 19, 1933. However, during the Great Depression, there was simply no market for the bonds for the project, and the Bureau of Reclamation took over construction. It had been envisioned that the state would operate and control the project once it was completed, but with the beginning of water deliveries from Shasta Dam in 1951, the attempt for state control over the CVP essentially ended. The principal transport of water by the project is from north to south. Whereas this southward transport can be accomplished via the Sacramento River north of the Delta, canals and pumping plants must be used to transport water from the Sacramento River to service areas south of the Delta (see figure 4-2).

Farming interests that opposed the 160-acre ownership limitation and the residency requirements on federal projects argued for state construction of any future water facilities. Their efforts resulted in the State Water Resources Development Bond Act of 1959, providing for financing of a State Water Project, subject to ratification by the voters. The MWD opposed the State Water Project and announced plans to develop the Eel River in northern California. This gesture of defiance prompted memories throughout the state of Los Angeles' activities in Owens Valley, and some southern coastal communities began independently endorsing the state project. The MWD eventually dropped its opposition, and the bond issue passed (Kahrl, 1978).

The SWP consists of a dam at Oroville (70 miles north of Sacramento) to store water from the Feather River for augmentation of the Sacramento River during dry periods, and delivery systems from the Delta for supplying water to the San Francisco Bay area, the San Joaquin Valley, and urban areas of the southern California coastal plain (see figure 4-2). Table 4-1 shows the supplies expected from the existing facilities of the SWP for the years 1985 to 2000 under the three hydrologic conditions postulated by MWD. These supplies show a slight decrease under average-year conditions by the year 2000. Sustained dry-period conditions or a severe short-term drought would severely reduce the supplies obtainable by MWD from the state.

Total Supplies and Estimated Shortfalls

Table 4-1 compares MWD's projections of demand with MWD's estimates of the total supply available within the service area from existing facilities. Under average conditions, the supply is projected to drop by about 80,000 acre-feet by the year 2000 because of additional Indian claims, use of water by desert power plants, and a slight reduction in SWP supplies. Demand in the MWD service area is projected to rise from 3.19 maf in 1985 to 3.61 maf in the year 2000. These estimates would leave a deficit of 140,000 acre-feet by the year 2000 under average water conditions and deficits as large as 630,000 acre-feet in a prolonged dry period, such as the period from 1928 to 1934. A repeat of the 1976/77 drought conditions could lead to a shortage of 610,000 acre-feet if it occurred in 1985 or of 1.2 maf if it occurred in the year 2000. Succeeding sections of this chapter will be concerned with various alternatives for balancing demand and supply. Many of these alternatives are currently being considered by MWD.

Alternatives for Balancing Supply and Demand

Additions to the State Water Project

MWD publications carry the message that additional SWP storage is an essential component of any long-term solution to its water supply needs (MWD, 1983b). However, California voters rejected construction of the Peripheral Canal across the Delta, which would have been the least expensive addition on a dollar-per-acre-foot basis to the SWP. Table 4-4 lists four other potential additions to the SWP and one potential addition to the federal CVP, as well as their expected yields and costs per acre-foot. The yields from future SWP facilities shown in table 4-4 would have to be allocated among MWD and other state contractors. The MWD expects to receive about 50 percent of the additional yields.

The Cottonwood Creek Project and the Thomes-Newville Reservoir would be located north of the Delta near Red Bluff. These reservoirs could provide an increase in firm yield even without a cross-Delta facility. After adding energy costs for delivery to MWD and facility costs,

TABLE 4-4. Potential Future Yields from State Water Project and Central Valley Project[a]

Project	Firm yield (maf)		Cost (1981$ per acre-foot)		
	With Delta transfer	Without Delta transfer	Costs at Delta	Transport to southern California	Total
State					
(Peripheral Canal)	0.70	—	100	110	210
Cottonwood Creek	0.20	0.15	200	110	310
Thomes-Newville	0.22	0.16	245	110	355
Los Vaqueros	0.26	0.20	325	110	435
Los Baños Grandes	0.25	0.10[b]	330	110	440
Federal					
Enlarged Shasta	1.40	[c]	175	110	285
Surplus water	1.00+	1.00	[d]	110	[d]

Sources: California Department of Water Resources, *State Water Project—Status of Water Conservation and Water Supply Augmentation Plans*, Bulletin 76-81 (Sacramento, Calif., 1981) pp. 30–35, table 3, and p. 36, table 4; Metropolitan Water District of Southern California, *Water Supply Available to Metropolitan Water District Prior to Year 2000*, Report No. 948 (Los Angeles, Calif., 1983) pp. IV-14 to IV-20.

[a] MWD would receive about half of the yields shown for the State Water Project but currently has no contracts to receive federal CVP water.

[b] Estimated yield if operated in conjunction with Los Vaqueros Reservoir.

[c] Not available.

[d] Although not available, this cost is estimated to be somewhat less than the cost of enlarging Shasta Dam.

water from Cottonwood Creek would have the lowest unit cost of the state facilities, with an incremental cost of some $310 per acre-foot. The Los Vaqueros Reservoir would be an offstream reservoir located south of the Delta. With the addition of a new pumping plant adjacent to the existing Delta Pumping Plant (but without a cross-Delta facility), the reservoir would be capable of storing water diverted from the Delta during the winter months. Another reservoir site capable of providing carryover storage would be the Los Baños Grandes site, just south of the San Luis Reservoir. The per-acre-foot costs of these last two facilities would be considerably higher than for Cottonwood Creek and Thomes-Newville.

There are several points of interconnection between the state and federal systems: the Sacramento River, the Sacramento-San Joaquin Delta, and the San Luis Reservoir (see figure 4-2). Therefore, an addition to either system could, in principle, provide additional water supply to the southern California coastal plain. An enlargement of Shasta Dam could provide water at a cost of $285 per acre-foot, some $25 per acre-foot cheaper than the most economically attractive of the proposed state facilities. The existing Shasta reservoir provides water for federal CVP contractors, and the sharing of any additional yield between the state and federal governments would need to be negotiated. Currently, excess capacity exists in the California Aqueduct of the SWP, and this capacity could be used to deliver water from an enlarged Shasta Dam.

Purchase of Surplus Water from the Federal Central Valley Project

Another way that the SWP might obtain water from the federal project is through purchase of existing supplies not now used by CVP customers. This amount could be augmented by completion of a coordinated operating agreement currently being negotiated between the state and the Bureau of Reclamation. In addition to specifying the amount of water to be released from state and federal reservoirs for water quality control in the Delta, the agreement is designed to improve the efficiency of joint operations. The surplus in the federal project after completion of the coordinated operating agreement is expected to exceed 1.0 maf. It may be possible for the state to purchase some of this surplus for its contractors, at least on an interim basis.

Demand Management

Demand Management Through Pricing. The MWD projects urban water demand by projecting future population growth and future per-capita use for each of its member agencies (MWD, 1982). The projections of per-capita use are based on an extrapolation of historical trends, combined with professional judgment. For example, in Burbank, Pasadena, and Long Beach, per-capita water use reached a maximum in 1972 and has decreased since then. The MWD believes the reduction, which they expect to continue, is due to several factors (MWD, 1982): public response to water conservation programs; laws enacted in recent years mandating the installation of low-water-use plumbing fixtures in new construction; replacement of single-family homes by multiple-dwelling units with reduced yard and garden water demands; and the imposition of sewer discharge fees by sanitation agencies in the early 1970s. These discharge fees cover part of the cost of sewage treatment and provide an incentive for industry to reduce water use by recycling or other means (the remaining sewer costs are covered by property tax assessments).

An alternative way for MWD to estimate future water demands would be to use estimating equations that specify future per-capita use as a function of various indicators, including lot size, income level, average temperature, sewer discharge fees, and industrial concentration. Using statistical techniques to relate these factors to per-capita water use would render future projections more sensitive to various underlying influences. In particular, statistical techniques would allow an estimation of the effect that the future price of water would have on the quantity of water demanded, an effect not now considered by MWD in its projections. The MWD recently commissioned a consulting engineer to study the price elasticity of demand for urban water (Hildebrand, 1984). In summing up the report, the general manager of MWD stated:

> a convincing demonstration has not been made for the existence of price elasticity of urban water demand; and if it exists, its effect does not appear to be as great as generally claimed. . . . It is concluded that no specific change should be made in projections due to price elasticity of water used for urban purposes (MWD, 1984).

This conclusion is not consistent with local experience with sewer discharge fees or with findings of statistically significant elasticities by at least three investigators of urban water supply in southern California: -1.02 to -1.09 (Conley, 1967), -0.31 (Gershon, 1968), and -0.26 to -0.41 (Schelhorse and coauthors, 1974).

Using these studies, some estimates can be provided of the effects that pricing policies would have. The MWD does not have long-term contracts with its customers; rather, each agency submits a revised list of projected demands annually, and this demand is subject to adjustment during the year. The MWD has some of the highest priced water available and is the residual source after local supplies are used. Therefore, MWD can be regarded as the marginal supplier for its member agencies. Likewise, additions to the SWP can be regarded as reflecting the long-run marginal cost of supply to MWD. SWP rates are being increased from $122 per acre-foot for 1982 to $221 per acre-foot for 1985 (in constant 1981 dollars), an increase of $99 per acre-foot (California Department of Water Resources, 1982, 1983). Therefore, if final consumers were charged rates based on SWP charges, they would experience rate increases of at least $99 per acre-foot.[1] Using $250 per acre-foot as a representative 1981 rate to consumers (1981 rates to final consumers in the MWD service area typically range between $200 and $300 per acre-foot once local distribution and treatment costs are added) and using an elasticity of -0.26 (the value at the lower end of the range in the studies of urban water demand in southern California), a $99-per-acre-foot rate increase would eventually reduce the quantity demanded by 300,000 acre-feet.[2] This volume of water is 71 percent of MWD's estimated increase in service area demand between 1985 and 2000 (420,000 acre-feet—refer to table 4-1) and exceeds the projected deficit under average-year conditions (140,000 acre-feet).

If consumers faced a water rate based on the cost of the least expensive of the proposed additions to the SWP ($310 per acre-foot), further reductions in demand would be expected. The $188-per-acre-foot increase in price would, with an elasticity of -0.26, reduce the quantity of water demanded by 490,000 acre-feet. This reduction exceeds MWD's predicted increase in service area demand between 1985 and 2000. Therefore, this estimated reduction indicates that there may not be enough water users in the MWD service area that value water highly enough to make an addition to the SWP a good economic investment.

Using an elasticity of -0.26 yields conservative estimates of the reductions that would occur under different pricing policies. If an elasticity of -0.40 is used, a value closer to the middle of the range of estimates for southern California and typical of residential use nationally (Howe and Linaweaver, 1967), then charges to final consumers based on SWP rates would reduce the quantity of water demanded by at least 450,000 acre-feet, and charges based on SWP additions would reduce the quan-

---

[1]The actual increase would be more because (1) MWD receives less-expensive water through the Colorado River Aqueduct and bases its charges on average cost, and (2) most jurisdictions within MWD currently cover a portion of their water and sewer costs by means of property tax assessments rather than through commodity charges for water and for sewage disposal.

[2]The constant elasticity demand equation (based on a year 2000 demand of 3.61 maf and a price of $250 per acre-foot) used to derive this estimate is $q = 15.17p^{-0.26}$.

tity demanded by 730,000 acre-feet.[3] Both of these amounts exceed MWD's projected increase in service area demand between 1985 and 2000.

In addition to consideration of the long-run marginal costs of supply, the use of marginal-cost pricing as a demand-management tool in the MWD service area would need to take into account how daily and seasonal peak use strain the capacity of different parts of the storage and delivery systems. Implementing higher "peak" rates during periods of increased use could serve to spread demand to off-peak times, as well as ensuring that those water users demanding peak service paid for the cost of increasing capacity. Prices based on marginal costs would also need to vary within the service area to reflect differences in local costs of storage and conveyance.

Demand Management in Droughts. MWD's water-supply objective is to fully satisfy its normal water demand projections under all supply conditions. Given the increasing expense of water supply facilities, drought management (that is, supplying less than normal demand) may be an attractive alternative to maintaining a full level of supplies during drought periods. In fact, MWD has practiced drought management in the past. The conservation programs adopted by MWD during the 1976/77 drought reduced total water use in MWD's service area by 14 percent (MWD, 1982). The same percentage reduction would lower demands in the year 2000 by 505,000 acre-feet, which is 42 percent of the deficit projected by MWD for the year 2000 under 1976/77 drought conditions and 80 percent of the deficit projected for the year 2000 under 1928–1934 drought conditions (refer to table 4-1). Considering the infrequency of drought conditions of these magnitudes and reductions in water use of no more than 14 percent, managing drought with some measure of planned reductions in water use may be one component in cost-effective water-supply planning.

Water Conserved by Investment in the Imperial Irrigation District

An examination of water use by the Imperial Irrigation District (IID) reveals why conservation measures in the district might be a viable source of water for MWD. First, the district has consumed up to about 3 maf annually of Colorado River flows, which is about 60 percent of California's past consumptive use of the Colorado and about 25 percent

[3]The constant elasticity demand equation used to derive these estimates is $q = 32.86p^{-0.40}$.

of the total annual net diversions from the river, as indicated in table 4-3. Second, an enormous amount of "plumbing" is required to deliver this water to some 450,000 irrigated acres: 1,627 miles of main canals and laterals; 1,454 miles of main drains and lateral drains; three regulating reservoirs; and numerous gates, checks, wasteways, spillways, and customer turnouts. In addition, the district manages 80 miles of the earth-lined All-American Canal. The customers are responsible for their own ditches, field siphons, holding basins, and some 26,000 miles of onfarm tile drains. Seepage and operational spills in a system of this size can add up to a significant amount of water.

Another consideration that constrains the efficiency of water use is the limited ability to manage deliveries once orders are placed. Because the capacity of the district's own regulatory reservoirs is relatively limited, the district must essentially order water 6 to 11 days in advance from Hoover Dam on the Colorado River (DWR, 1981a). If farmers change their demands for water while the water is in transit, the district must try to shuffle the deliveries to another user, leave them stored in district canals, or spill them into wasteways.

A further indication that there may be additional potential for conservation is the relatively low price that IID farmers pay for water. Under provisions of the Boulder Canyon Act, IID pays nothing toward the construction cost of Hoover Dam, but does repay $25,020,000 of the construction cost of the All-American Canal on the interest-free basis accorded agriculture under Reclamation law (U.S. Department of the Interior, 1977). The current federal payment required of the district is $750,000 per year, regardless of the amount of water taken. This amount averages about $0.25 per acre-foot, but the district is responsible for operation and maintenance and adds its own charges to the federal repayment. The district currently charges $9 per acre-foot for water deliveries (IID, 1983). In addition, farmers pay an assessment against lands of $4 per acre. The $9-per-acre-foot charge for irrigation water is an approximate measure of the incentive that farmers have to prevent excess use of water. This charge is far below the price that MWD would pay for conserved water.

The district itself has initiated several water conservation measures. The district began lining canals in 1960. In July 1976 the district formally adopted a thirteen-point water conservation program, and in 1980 it adopted a twenty-one point water conservation program (California Department of Water Resources, 1981a). These programs include structural measures (such as construction of an additional water-regulating reservoir and reconstruction of farm outlet boxes), price incentives (such as providing free drainage water to persons willing to pump it, and charging users who waste water an assessment equal to three times the scheduled water rate), more flexible delivery of water off-schedule, and initiation of an irrigation management services program.

These programs were not successful enough to prevent one of the district's farmers, John J. Elmore, from filing a complaint with the state in June 1980 alleging misuse of water by the district. Provisions of the California Constitution calling for the prevention of "waste or unreasonable method of use of water" (Article 10, Section 2) are implemented by Sections 100 and 275 of the California Water Code, which require the Department of Water Resources and the State Water Resources Control Board to investigate alleged misuse. Elmore complained that inefficient practices by the district were creating waste flows to the Salton Sea that were threatening to inundate his property. He had already found it necessary, at his own expense, to dike his farmland and to install pumps to remove excess water. Among the district practices cited by Elmore were the absence of sufficient regulatory reservoirs, the inflexibility of deliveries (water must be ordered in 24-hour increments), and an insufficient number of tailwater recovery systems.

In December 1981 the California Department of Water Resources published its findings from the investigation of Elmore's complaint, setting forth the estimated losses of water, the quantities of water that could be conserved in the district by various measures, and the estimated costs of such measures (DWR, 1981a). As indicated in table 4-5, estimated annual losses total 901,000 acre-feet, or about a third of the water delivered to Imperial. The potential conservation measures are presented in the same table in approximate order from the least to the most costly measures per acre-foot. The estimated costs range from $14 per acre-foot for construction of additional spill-interceptor canals to $115 per acre-foot for lining the All-American Canal. The total quantity of water conservable, assuming investments through the upper range of costs, is estimated at 437,000 acre-feet.

The Bureau of Reclamation also has prepared a special report on water conservation opportunities in the Imperial Irrigation District (U.S. Department of the Interior, 1983), and estimates from the economic appendix to the report are reproduced in the last columns of table 4-5. The estimates are similar to those from the DWR report, although there are notable differences. The Bureau finds a larger quantity of water available through the construction of spill-interceptor canals, but a smaller quantity through onfarm management. The Bureau also places a higher cost on the construction of regulatory reservoirs and district canal lining. Figure 4-3 is a graph of the water-supply curve available from conservation based on the two studies.

To simplify the presentation of subsequent estimates, one set of conservation costs is used: the four summary increments in table 4-6 and

TABLE 4-5. Estimated Water Losses and Costs of Conservation Investments in Imperial Irrigation District (1981 dollars)

Investment	Department of Water Resources estimates of losses Quantity (acre-feet)	Department of Water Resources estimates of potential savings Cost (dollars per acre-foot)	Department of Water Resources estimates of potential savings Quantity (acre-feet)	Bureau of Reclamation estimates of potential savings Cost (dollars per acre-foot)	Bureau of Reclamation estimates of potential savings Quantity (acre-feet)
Spill-interceptor canals	78,000	14	30,000	14	70,000
Onfarm management	570,000				
Tailwater recovery		8–25	178,000	18	135,000
Other		varies			
Canal spills	53,000				
Flexible deliveries		27		—	—
System automation		unknown	50,000	27	25,000
Regulatory reservoirs		34		100	20,000
Line main canals	122,000	31	110,000	68	100,000
Line All-American Canal	73,000	115	70,000	112	51,000+
Total quantity	901,000		437,000		401,000+

Sources: California Department of Water Resources, *Investigation Under California Water Code Section 275 of Use of Water by Imperial Irrigation District,* Southern District Report (Los Angeles, Calif., 1981) p. 56, table 13; p. 59, figure 9; and pp. 60–61, table 15; U.S. Department of the Interior, Bureau of Reclamation. "Water Conservation Opportunities: Imperial Irrigation District, California," Draft Special Report (Washington, D.C., 1983) Economic Appendix, p. 24, table 4; Bureau of Reclamation estimates for lining the All-American Canal are taken from U.S. Department of the Interior, Water and Power Resources Service (Bureau of Reclamation). *Reject Stream Replacement Study,* Colorado River Basin Salinity Control Project, Special Report (Washington, D.C., 1980) p. 7, table 1. Costs are converted to 1981 dollars using the GNP price deflator.

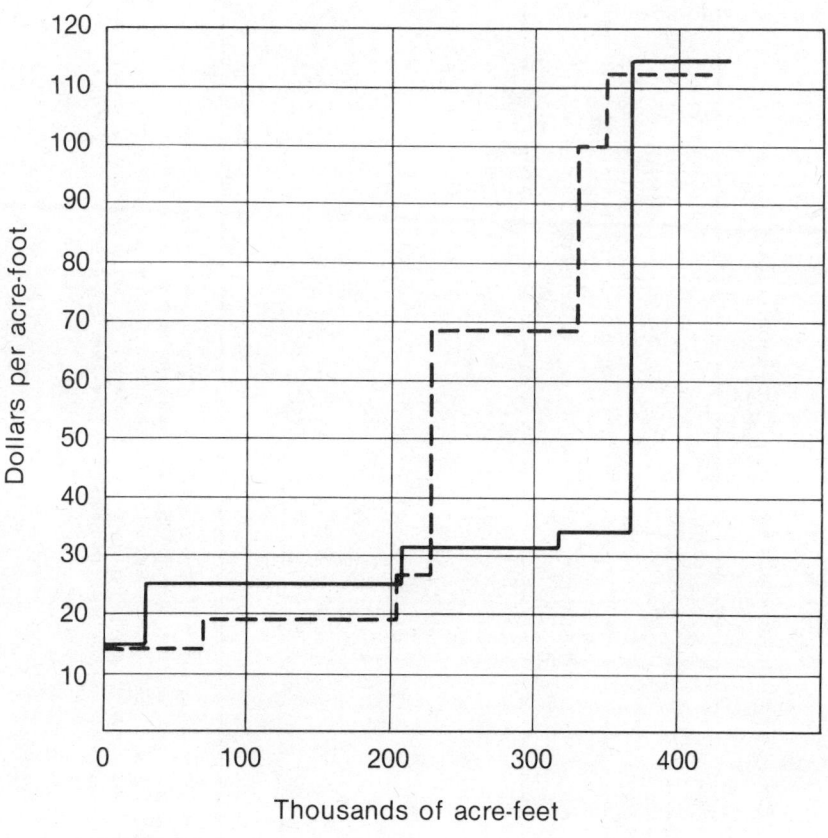

Figure 4-3. Water supply from conservation investments in the Imperial Irrigation District Source: Table 4-5

the higher of the estimates from figure 4-3. As table 4-6 shows, the total cost of conserving water in IID and delivering it to MWD would include not only the cost of the conservation investments themselves, but also delivery costs and certain other adjustments. Water conserved and sent through the Colorado River Aqueduct to the Los Angeles area would decrease power production at Parker Dam (see figure 4-1). Recent estimates are that from 60 to 66 kWh of electrical energy would be forgone for every acre-foot reduction at Parker Dam (EDF, 1983; U.S. Department of the Interior, 1983, Economic Appendix). Valuing the higher estimate at a daily average of on- and off-peak values ($0.0362

TABLE 4-6. Total Cost of Conserving Water in the Imperial Irrigation District and Delivery to the Metropolitan Water District

Quantity/cost	Units			
Quantity of water conserved (acre-feet)				
Total quantity	30,000	230,000	330,000	400,000
Incremental quantity	30,000	200,000	100,000	70,000
Marginal cost (dollars per acre-foot)				
Conservation costs	14	25	68	115
Energy forgone				
Parker Dam	2	2	2	2
(66 kWh/acre-foot)				
IID power plants	5	5	5	5
(134 kWh)/acre-foot)				
Colorado River Aqueduct				
Pumping	72	72	72	72
(2000 kWh/acre-foot)				
Salinity costs[a]	105	105	105	105
Total	199	210	253	300
Average cost (dollars per acre-foot)				
Conservation cost	14	19	36	55
Energy & salinity	185	185	185	185
Total	199	204	221	240

Sources: Conservation costs are from figure 4-3 and table 4-5. For sources of energy costs see text. For salinity costs see Environmental Defense Fund, *Trading Conservation Investments for Water* (Berkeley, Calif., Environmental Defense Fund, Inc., 1983) p. 67, table 15. Columns may not add due to rounding.

[a] Salinity costs are deflated from 1986 dollars at 9 percent per year (see Environmental Defense Fund, *Trading Conservation Investments for Water*, p. 68).

per kWh)[4] leads to an estimate of $2.39 per acre-foot as the cost of forgone energy.

Further energy loss would be incurred at energy facilities operated by IID itself. The district operates hydroelectric facilities at five "drops" along the All-American Canal, a facility at the East Highline Canal Turnout, and two smaller generators. Any reduction in diversions of water to the district would reduce the flows through these facilities, with the exception of the facility at Pilot Knob. Because water can be turned back into the Colorado River from Pilot Knob, the canal could be kept at its original level at this point with no loss of power production. Under the assumption that a reduction in diversions entering the system at Drop No. 1 would lead to proportional reductions throughout the system, the Environmental Defense Fund (EDF) estimated a total reduc-

[4]This energy value is based on Southern California Edison's 1982 estimates of replacement energy for 1986. The estimates are deflated to 1981 dollars at 8.9 percent per year (see U.S. Department of the Interior, "Water Conservation Opportunities: Imperial Irrigation District, California," Draft Special Report (Washington, D.C., 1983) Economic Appendix, p. 32, and Environmental Defense Fund, *Trading Conservation Investments for Water* (Berkeley, Calif., 1983) p. 73, note 19).

tion of 106.12 kWh per acre-foot (EDF, 1983). The district reports power generation to be approximately 134 kWh per acre-foot (U.S. Department of the Interior, 1983, Economic Appendix). Valuing this latter amount at $0.0362 per kWh leads to an additional $4.84-per-acre-foot cost that MWD would incur in order to compensate IID.

About 2,000 kWh per acre-foot is the net energy to pump water through the Colorado River Aqueduct. Table 4-6 shows that this cost adds another $72 per acre-foot to the cost of water diverted at Parker Dam. The MWD blends water from various sources, of which the Colorado River is the most saline. The greater salinity of the blended water than of an alternative supply from the SWP would impose costs on plumbing systems and household appliances in the MWD service area. Estimates by the California Department of Water Resources and by EDF (1983) are that about $105 per acre-foot in costs would result. These are not financial costs to MWD but costs that would be borne by consumers of the water.

Table 4-6 shows that the total marginal costs of water conserved in IID and delivered to MWD, including salinity costs, would range from $199 per acre-foot to $300 per acre-foot. The average cost of a full 400,000 acre-feet of conservation investments would be $240 per acre-foot. Therefore, water deliverable because of conservation investments in IID appears to be an attractive alternative.

There could also be some environmentally related costs of the diversion of water to MWD. Most notable among these is that the reduced inflows to the Salton Sea would lead to a lower lake elevation and increased salinity, endangering the fish population and reducing recreational benefits. However, indications are that salinity levels dangerous to fish are expected to be reached nearly as rapidly without conservation measures as with them. Perhaps the best indicator of the low level of environmental cost expected from the transfer is the Environmental Defense Fund's assessment (1983) that "any minor degradation that may occur as the result of more efficient water use is highly unlikely to serve as the basis for a valid legal claim barring those water savings."

Other Alternatives

Other means of balancing demand and supply are generally less certain, more difficult to quantify, or offer lesser quantities of water than the major alternatives considered above. Many of these additional alternatives are discussed by MWD in its planning reports (MWD, 1983b). Within MWD's service area, proposed storage facilities at Pamo Valley and on the Santa Margarita River would provide dependable supplies of about 11,000 acre-feet each and could each provide about ten times that capacity for storing flows obtained from the Colorado River or from the SWP. Storage of imported water in the Chino groundwater basin would be another means for providing supplemental supplies during prolonged dry periods. As of 1983, MWD's wastewater reclamation program provided about 63,000 acre-feet of reclaimed water. The MWD believes that each year an additional 5,000 acre-feet of reclaimed water can be obtained from future projects. In addition, it has marketed an interruptible class of water service. The MWD can choose not to deliver this water, currently amounting to 625,000 acre-feet, for a period of one year, thus reducing demands during a dry period.

Large quantities of water have seeped from the unlined portions of the Coachella and All-American canals. Should these canals be lined, this water could be pumped out and supplied to the irrigation districts in exchange for an equal supply of their water from the Colorado River. Another possibility is for surplus flows from the Colorado River to be stored as groundwater in the Coachella area and later pumped out for use in Coachella in exchange for increased Colorado River deliveries to MWD.

There are also some potential sources that are not discussed by MWD in its water-supply reports. The MWD places its conveyance losses from the Colorado River Aqueduct at 50,000 acre-feet per year. This is about 5 to 7 percent of recent aqueduct deliveries, suggesting that the feasibility of conservation measures should be investigated. Furthermore, some of MWD's estimates of other losses from its Colorado River entitlements may prove too pessimistic. Southern California Edison recently canceled the proposed 2,000-megawatt Allen-Warner plant, indicating that the construction of desert power plants may be delayed beyond the year 2000, thereby freeing another 42,000 acre-feet per year. Moreover, the Indian claims have yet to reach a settlement, and this introduces another element of uncertainty into the MWD estimates.

Conservation investments in the SWP or the CVP (similar to those proposed in IID) may be worth future consideration. Although the energy costs of delivering water to MWD would be an estimated $38 per acre-foot higher than from Imperial, this would be more than offset by $105-per-acre-foot savings to consumers due to reduced salinity. An evaluation of this alternative would need to take into account the value of return flows in the Sacramento and San Joaquin valleys.

Prospects for Economically Efficient Alternatives

Table 4-7 summarizes the estimates developed for the four basic alternatives for balancing future water supply and demand. A comparison

TABLE 4-7. Summary of Alternatives for Balancing Supply and Demand in Metropolitan Water Districts

Alternatives	Projected firm yield or demand reduction (maf)	Unit cost (1981 dollars per acre-foot)
State Water Project—No cross-Delta facility		
Cottonwood Creek	0.07	310
Thomes-Newville	0.08	355
Los Vaqueros	0.10	435
Los Baños Grandes	0.05	440
Federal Central Valley Project		
Enlarged Shasta Dam	a, b	285
Surplus water	1.00[b]	c
Demand management		
Through pricing in MWD		
at SWP rates	0.30 to 0.45	d
at SWP incremental costs	0.49 to 0.73	d
During droughts	0.50	d
Colorado River Contractors—Imperial		
Spill-interceptor canals	0.03	199
Tailwater recovery	0.20	210
Lining main canals	0.10	253
Lining All-American Canal	0.07	300

Sources: Tables 4-4 and 4-6. The yields shown here for the SWP reflect the fact that MWD expects to receive about 50 percent of the yields from the future SWP facilities shown in table 4-4. For sources of demand management estimates, see text.

[a] Not available. Yield with a cross-Delta facility is 1.40 maf.

[b] State purchase of water from federal facilities would need to recognize the expectations of federal contractors.

[c] Not available. The cost is estimated to be somewhat less than enlarging Shasta Dam.

[d] Not available.

of the supplies expected from these alternatives with the "normal demand" deficit of 140,000 acre-feet for the year 2000 (see table 4-1) indicates no cause for alarm. Various alternatives now under consideration would more than offset MWD's deficit projections. One economically efficient alternative would be to manage demand by increasing the price of water to approach marginal cost. Although consumers would be giving up the benefits of somewhat greater quantities of water, the benefits associated with the larger quantities would be less than the cost of providing the additional water. This is typically the case where marginal-cost pricing replaces average-cost pricing.

If MWD could obtain surplus supplies from the federal Central Valley Project at the current CVP average cost rates, this would be an inexpensive alternative. The Bureau of Reclamation's actual expenditures for these supplies, however, would probably be close to $285 per acre-foot, and the federal government might well require full payment of these costs.[5] Therefore, the alternative of investing in conservation measures in IID may be less expensive to MWD. The additions to the State Water Project appear to be the most expensive of the alternatives.

Three of the more economically efficient alternatives for meeting demand under normal hydrologic conditions are purchase of surplus supplies from the CVP, demand management through pricing, and conservation investments in IID.

Prospects for Purchase of Surplus Water from the Federal CVP

The MWD faces a legal constraint on the use of water from the CVP; namely, this water must be used within the authorized service area of the federal project, which does not extend far enough south to reach MWD. This constraint might be overcome, however, if the state purchased the water conserved in the CVP for SWP users. This would free additional water in the state system for delivery to MWD.

An offer by the state to purchase surplus water from the CVP might well encounter resistance from the Bureau of Reclamation, which is accustomed to allocating water to districts with which the Bureau has already established contractual relationships. In fact, the State of California wrote the U.S. Department of the Interior in 1981 offering to purchase surplus water. The state was reportedly willing to pay an amount close to the full incremental cost of recent additions to the federal project although this was not an explicit part of the offer. The offer was made near the end of the Carter administration and was never acted upon. However, growing demand for water in the state may bring increasing pressure for the federal government to sell or lease some of its current surplus to the state project. Any such agreement would need to recognize, in some form, the expectations of agricultural users of federal water supplies.

Prospects for Demand Management Through Pricing

Demand management through pricing would face several hurdles. The first is that price increases can be difficult to put in place and are never

---

[5]Since the construction cost is a sunk cost, $285 per acre-foot is not the economic value of the water. Allocating the surplus water to municipal and industrial uses would improve economic efficiency as long as users valued the water more than the value in its current uses—interim contracts and water quality improvements in the Delta.

popular with ratepayers. Still, an increase in water rates may be more easily understood by the public at this time because the SWP is raising its rates to southern California. If MWD and its member agencies were to price water based on the most expensive increments to supply, they would face another problem: the revenues collected would far exceed costs, thereby violating their revenue constraint as public utilities. This obstacle could be overcome by eliminating the current property tax assessments for water supply and sewage treatment and by using the estimated surplus revenues to further reduce property taxes. This offset against property taxes might also help reduce public resistance to increased water rates.

A further obstacle to full implementation of marginal-cost pricing is the degree of cooperation required among MWD's twenty-seven member agencies. If MWD charged a rate based on marginal cost to its members, but the agencies did not price at marginal cost, the effect of marginal-cost pricing by MWD would be diminished. Complete implementation of a marginal-cost pricing scheme would require that marginal costs be carried by the member agencies down to the level of the final consumer. To reach this degree of consensus is likely to require considerable time. A first step would be for MWD to analyze the demand, revenue, and property tax implications of using different pricing and taxing policies in its service area.

Prospects for an IID-MWD Transfer

Legal Constraints. The first question that arises with regard to the possible transfer of water from the IID to the MWD is whether under Reclamation Law, the Imperial Irrigation District has sufficient property rights in its Colorado River water that it would be allowed to sell or transfer water to MWD. The answer appears to be yes. The district, rather than the Bureau of Reclamation, holds the water rights. These rights predate any federal allocation of additional water from Hoover Dam, and these rights were recognized by the Colorado River Compact, the Boulder Canyon Project Act, and the Supreme Court's decision in *Arizona* v. *California* (Nathanson, 1978).

A major tenet of appropriative water rights doctrine is that water must be put to a beneficial use by the holder of the right in order for the right to prevail. Therefore, a second question is whether Imperial can lease, sell, or otherwise transfer a portion of its water to MWD without losing the rights to that portion. Various western states have recognized the negative effect that the original definition of beneficial use has had on water conservation: namely, water users have no incentive to finance conservation investments if the water conserved is to be shifted, without compensation, to junior appropriators. For this reason, certain states, including California, have moved to define the conservation and transfer of water as a beneficial use. In 1982 California took an additional step when a bill sponsored by Congressmen Katz and Bates became law and amended certain sections of the California water code. The conservation concept was strengthened by allowing a water agency to "sell, lease, exchange, or otherwise transfer water that is surplus to the needs of the agency's water users for use outside the agency." "Surplus" was broadly defined and includes water that an agency and a water user agree to forgo using during the period of the transfer. Although the amendments to the code restrict certain transfers to seven years' duration, this restriction does not apply to water developed through wastewater reclamation or water conservation.

There is also a question whether the Coachella Valley County Water District would have the rights to any water conserved by Imperial. Coachella and Imperial both have third priority for water use under the Seven Party Agreement, whereas MWD has fourth priority (table 4-2). The issue of how water would be divided between Imperial and Coachella came up soon after the Seven Party Agreement was established. In 1934 Imperial and Coachella signed a compromise agreement (U.S. Department of the Interior, 1950) in which Imperial was given the prior right for water over Coachella. Therefore, under California law it appears that Imperial might be able to lease or sell water at a profit to MWD without interference.

Competing Uses for the Conserved Water. Apart from the legality of the transfer to MWD is the question of whether Imperial or Coachella might desire to put the conserved water to use themselves instead of transferring it to MWD. The compromise agreement limits either district from adding more than 5,000 acres to its service area without the written consent of the other, as well as the consent of the Secretary of the Interior. Therefore, Imperial would be restricted from using significant quantities of water to expand district acreage unilaterally. Furthermore, despite occasional protestations to the contrary, there is evidence that neither district wants to expand its acreage. When the lining of the first 49 miles of the Coachella Canal took place, Imperial had the opportunity to pay for additional capacity in the canal to irrigate its East Mesa lands. Imperial declined. Some Imperial farmers recognize that to expand the acreage of the district would be to increase competition for their own produce. The attitude toward expansion on the sandy mesa lands is reflected in the fact that the district has established a special water rate schedule for these lands. The district levees double the rate for water in excess of 6 acre-feet per acre and quadruple the rate for water in excess of 8 acre-feet per acre (IID, 1983).

Central to the question of whether the district would seek to apply the conserved water to expanded acreage rather than leasing it to MWD is the economic benefit of the water conserved in agricultural use relative to municipal and industrial uses. The Bureau of Reclamation (U.S. Department of the Interior, 1983) estimates that the economic benefit of irrigation would be approximately $35 per acre-foot on lands in the Imperial Valley and, because of the extra cost required to pump water up to the mesa, no more than $30 per acre-foot on the district's mesa lands. The costs of MWD's other alternatives would indicate that these values are far below the amounts MWD would be willing to pay for water.

State and Federal Support. Concerns about whether federal and state agencies would oppose an IID-MWD exchange seem to be unfounded. In a March 1984 response to a request from Congressman Vic Fazio of California regarding the transfer proposal, Secretary of the Interior William Clark indicated that there appeared to be no legal obstacle to the transfer and that it most likely did not require congressional approval. Governor George Deukmejian endorsed the proposal in a special message to the state legislature in April 1984.

On June 21, 1984, the California State Water Resources Control Board (SWRCB) rendered its decision on John Elmore's charges against Imperial (SWRCB, 1984). The board agreed that wastewater flows from Imperial were indeed contributing to rising lake levels in the Salton Sea and found that the practices set out in the district's thirteen-point and twenty-one-point conservation programs have been less than successful. The board's decision represents a middle ground in that it found Imperial's use of water "unreasonable," but not "wasteful." The order requires the district to (1) submit evidence that it has fully implemented its tailwater monitoring program; (2) repair, or require water users to repair, defective tailwater structures; (3) submit a detailed and comprehensive water conservation plan; and (4) submit a plan for resumption of the program for construction of regulatory reservoirs, including a proposed method for financing their construction. The board's decision probably makes an exchange with MWD more likely because it establishes deadlines for some of the measures to be taken, because IID will need to find a use for the conserved water that will avoid further increases in the level of the Salton Sea, and because the district would almost undoubtedly have to rely on outside financing for some of the more expensive measures to be required, such as the construction of additional regulatory reservoirs.

Negotiating the Transfer. A functioning market for water could probably have avoided the direct regulation of district activities implicit in the decision of the State Water Resources Control Board. Likewise, there are several ways that market processes might preserve flexibility for the negotiating parties as they work out the terms of a transfer. For example, IID could sell or lease water to MWD for a profit sufficient to allow IID to make conservation investments on its own. This approach would give IID more freedom to manage affairs within the district than would a cooperative work program under which MWD would construct the conservation investments.

The MWD will naturally be concerned about the long-term security of any water supplies it obtains, while IID will be reluctant to undertake any action that could lead to a permanent loss of rights to water that it could put to use at some time in the future. One way to satisfy both concerns would be for IID to grant a rolling long-term lease for the conserved water, say a twenty-year lease subject to twenty-year extensions after every ten years. This arrangement would allow MWD the security of a twenty-year supply and would provide a signal ten years in advance if the lease were not to be renewed. Correspondingly, IID would have only to wait out the term of the lease to regain any or all of the conserved water for its own use.

Furthermore, a market, with the district acting as broker for its farmers, might well be able to respond to changing conditions and changing demands for water more rapidly than cooperative work arrangements or state actions. For example, during a severe drought in California, some farmers in the IID might be willing to forgo production on marginal lands in return for compensation from MWD. Such a program in IID would allow MWD to purchase temporary increases in its deliveries from the Colorado River while reducing its deliveries from northern California, much as it did during the 1976/77 drought.

In 1984, the governing boards of IID and MWD initiated mutual discussions of a water transfer. Given the history of Los Angeles' dealings in the Owens Valley and the economic and political power of MWD, one could expect IID to proceed cautiously. Yet, there are many reasons to expect that the current situation is significantly different from that in the Owens Valley. The MWD has gone out of its way to convey a cooperative, rather than an aggressive, image. The MWD presented no evidence to the State Water Resources Control Board concerning possible waste of water by Imperial. Rather, MWD submitted a two-page statement indicating its willingness to cooperate with IID in a water-salvage program but emphasizing that "the agricultural and water future of the Imperial Valley can only be decided by the local people" (MWD, 1983a). This cooperative approach has proved fruitful: in 1985, MWD and IID reached a tentative agreement that MWD will provide $10 million annually toward conservation investments in IID in exchange for 100,000 acre-feet of water.

The contrast between current prospects and past practices is striking. Completion of a mutually satisfactory transfer between IID and MWD could go a great distance toward improving acceptance of such practices among the water-supplying regions of the state and could also establish a precedent for the success of future voluntary market transfers involving federally supplied water.

References

California Department of Water Resources (DWR). 1981a. *Investigation Under California Water Code Section 275 of Use of Water by Imperial Irrigation District.* Southern District Report (Los Angeles, Calif.).
——. 1981b. *State Water Project—Status of Water Conservation and Water Supply Augmentation Plans.* Bulletin 76-81 (Sacramento, Calif.).
——. 1982. *The California State Water Project—Current Activities and Future Management Plans.* Bulletin 132-82 (Sacramento, Calif.).
——. 1983. *Management of the California State Water Project.* Bulletin 132-83 (Sacramento, Calif.).
California State Water Resources Control Board (SWRCB). 1984. *Imperial Irrigation District—Alleged Waste and Unreasonable Use of Water: Water Rights Decision 1600* (Sacramento, Calif.).
Colorado River Board of California. 1964. *Annual Report, 1963–64* (Los Angeles, Calif.).
Conley, Brian C. 1967. "Price Elasticity of the Demand for Water in Southern California," *Annals of Regional Science*, vol. 1, pp. 180–189.
Environmental Defense Fund (EDF). 1983. *Trading Conservation Investments for Water* (Berkeley, Calif., Environmental Defense Fund, Inc.).
Gershon, Sam I. 1968. *Unit Water Use Model for the South Coastal Area* (Los Angeles: California Department of Water Resources, Southern District).
Hildebrand, Carver W. 1984. "The Relationship Between Urban Water Demand and the Price of Water," The Metropolitan Water District of Southern California (Los Angeles, Calif.).
Hoffman, Abraham. 1981. *Vision or Villainy: Origins of the Owens Valley—Los Angeles Water Controversy* (College Station, Tex., Texas A&M University Press).
Howe, Charles W., and F. P. Linaweaver 1967. "The Impact of Price on Residential Water Demand and Its Relation to System Design and Price Structure," *Water Resources Research*, vol. 3, no. 1, pp. 13–32.
Imperial Irrigation District (IID). 1983. "Water Rate Schedules Nos. 1–7" (El Centro, Calif.).

Kahrl, William L., ed. 1978. *The California Water Atlas*. California Office of Planning and Research and Department of Water Resources (Sacramento, Calif.).

———. 1982. *Water and Power: The Conflict over Los Angeles' Water Supply in the Owens Valley* (Berkeley, Calif., University of California).

Metropolitan Water District of Sourthern California (MWD). 1982. *1982 Population and Water Demand Study*. Report No. 946 (Los Angeles, Calif.).

———. 1983a. "Statement of the Metropolitan Water District of Southern California Regarding Proceedings Alleging Misuse of Water by Imperial Irrigation District." Presented to the State Water Resources Control Board (Los Angeles, Calif.).

———. 1983b. *Water Supply Available to Metropolitan Water District Prior to Year 2000*. Report No. 948 (Los Angeles, Calif.).

———. 1984. Memorandum from General Manager to Board of Directors concerning "Transmittal of Report on Price Elasticity of Urban Water Demand" (Los Angeles, Calif.).

Nadeau, Remi A. 1950. *The Water Seekers* (Garden City, N.Y., Doubleday & Company).

Nathanson, Milton N., ed. 1978. *Updating the Hoover Dam Documents*, U.S. Department of the Interior, Bureau of Reclamation (Washington, GPO).

Schelhorse, Larry D., Peggy Zimmerman, Jerome W. Milliman, David L. Shapiro, and Louis F. Weschler. 1974. *The Market Structure of the Southern California Water Industry*. Prepared for the Office of Water Resources Research, U.S. Department of the Interior (La Jolla, Calif., Copley International Corporation).

U.S. Department of the Interior, Bureau of Reclamation (USBR). 1950. *Hoover Dam Power and Water Contracts and Related Data* (Washington, D.C.).

———. 1977. *Water & Land Resource Accomplishments, Project Data, Statistical Appendix III* (Washington, D.C.).

———. 1980. *Reject Stream Replacement Study*, Colorado River Basin Salinity Control Project, Special Report (Washington, D.C.).

———. 1983. "Water Conservation Opportunities: Imperial Irrigation District, California," Draft Special Report (Washington, D.C.).

5
Costs of Water Management Institutions: The Case of Southeastern Virginia

*Leonard Shabman and William E. Cox**

Conflicts over water use in arid areas are understandable if increased water consumption by one user may require reduced consumption by others. In Virginia, this type of water reallocation has not been necessary and appears unlikely in the future. Except for a few areas of the state, increased use of water for irrigation is limited by the questionable profitability of irrigation investment. Industrial and steam-electric power production do require large quantities of water, but a trend toward water recycling and reduced water use in production processes will reduce future increases in withdrawals. Increased urban water use can be expected, but relative to available streamflow this use is a modest one (Shabman and coauthors, 1981).

When compared with projected use, Virginia has an abundance of water. The average rainfall of more than 40 inches is well distributed throughout the year. Moreover, the state has extensive groundwater storage areas, especially in the eastern coastal plain. Despite this abundance, there has been intense public debate over proposals to transfer additional water across local political boundaries into some of the state's urban areas. In northern Virginia, after years of disagreement among political jurisdictions, an adequate future water supply has been assured by adoption of a plan for coordinating the region's separate water systems (Sheer, 1983). In the western area of the state, the city of Bedford

---

*Respectively, Department of Agricultural Economics, and Department of Civil Engineering at Virginia Polytechnic Institute and State University.

and Bedford County have been involved in a dispute over the city's proposal to drill wells in the county. In eastern Virginia, counties within the Pamunkey and Mattaponi river basins have resisted a proposal by the city of Newport News to transfer water from those river systems. The most sustained jurisdictional conflict has occurred over providing water to an urban area in southeastern Virginia, where several different proposals were made to transfer water to the cities of Norfolk, Chesapeake, and Virginia Beach, including one to take water from as far away as Lake Gaston in central Virginia.

Because of its abundance relative to future use, and its widespread distribution, water in Virginia is seldom an economically scarce resource. Therefore, the expansion of urban water systems need not be a source of conflict. However, because the need to transfer water has not been significant in Virginia's history, water allocation law and government management programs have developed limited capacity to resolve conflict, especially where political jurisdictions are involved.

These institutional inadequacies can reduce the efficiency of urban water management. One source of inefficiency is capital investment greater than that needed to provide a given amount of water and avoid water-use conflict. Another result of institutional inadequacy may be increased transactions costs for conflict resolution (for example, for legal fees and construction delays). In Virginia, however, inefficiency will rarely result from uncompensated opportunity costs imposed by urban water development on other users, because the abundance of water makes its scarcity value nearly zero.

In the section that follows, the water supply situation in southeastern Virginia will be described to illustrate how economic inefficiency may result from institutional inadequacies in the Virginia system. Conclusions will then be drawn about appropriate institutional reforms for the Virginia "water problem." (Because these institutional reforms to address potential inefficiencies are not without cost, they must be appropriate to the nature and extent of the problem being addressed.)

The Southeastern Virginia Water Situation

Water Supply and Use

The southeastern Virginia region for water supply planning extends from Virginia Beach on the east to Isle of Wight and Southampton on the west (see figure 5-1). The northern border is the James River, and the southern border is North Carolina. This geographic area has been treated as one water supply region because of the interconnected service

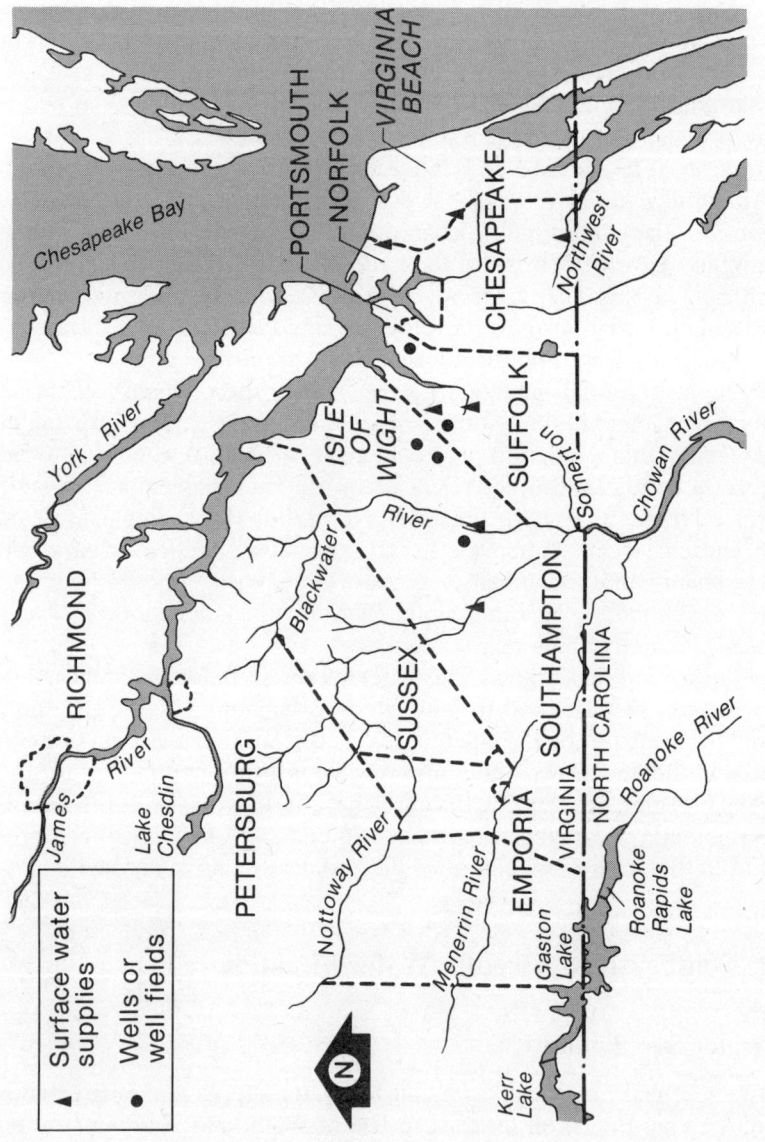

Figure 5-1. The study area

areas of the municipal water systems and because water sources in the western part of the region serve the easternmost cities.

In the last decade there have been numerous studies of the southeastern Virginia water supply situation, focusing on the cities of Norfolk, Virginia Beach, Chesapeake, and Portsmouth. In each of these studies, projections of water use for a fifty-year period were first compared with estimates of the yield of the existing water systems to determine future water deficits. Next, alternatives capable of eliminating the deficit by a specified time were determined and their feasibility analyzed.

In this period there have been differing estimates of the regional water deficit. During 1983 there was a convergence of studies on water use projections and water system yields, which have been incorporated into proposals for solving the area's urban water supply problems. The solution currently favored by the regional governments, with the support of the state, is construction of a pipeline to transfer water from Lake Gaston, in the central part of Virginia.

Water Use Projections

The Norfolk District of the U.S. Army Corps of Engineers made water use projections for the region as part of a congressionally authorized study (U.S. Senate Committee on Public Works, 1974). No other agencies have made projections, although local and state governments were consulted as part of the Corps' effort. However, projections varied over time in response to changes in federal planning procedures.[1] Through 1977, water use estimates were based on an extrapolation from historical trends. The 1977 projections of water use for the major political jurisdictions in the region are shown in panel A of table 5-1.

By 1978, new federal planning guidelines stressing reduced municipal water demand and improved management of water distribution were required for Corps of Engineers studies (Bauman and coauthors, 1980). Projections were to be based on the following assumptions: (1) water-saving techniques would be used in new homes and businesses, (2) water use in older homes and businesses would be partially reduced by retrofitting plumbing with water saving devices, (3) education programs and pricing policy would be in place to reduce water use, (4) changing industrial production practices would reduce water withdrawals by that

[1]Throughout the planning process, water use has been described in terms of average day demand. In this region seasonal variations in water use are not large (Anderson, 1978). Therefore, the need to consider peak-season demand in water planning was not considered in the planning process.

TABLE 5-1. Projections of Water Use for Southeastern Virginia (million gallons per day)

City	1980	1990	2000	2010	2020	2030
			A: 1977 projections[a]			
Norfolk	46.6	49.8	52.7	55.1	56.8	58.3
Virginia Beach	22.1	33.7	44.4	53.6	63.7	74.3
Chesapeake	8.7	12.7	16.7	20.7	24.7	28.7
Portsmouth	15.5	17.3	18.9	20.2	21.5	22.5
Total	92.9	113.5	132.7	149.6	166.7	183.8
			B: 1981 projections[b]			
Norfolk	44.35	43.08	43.61	43.67	43.94	44.62
Virginia Beach	22.50	31.02	38.20	42.59	47.21	52.14
Chesapeake	8.72	11.55	13.78	15.60	17.46	19.74
Portsmouth	14.53	15.41	16.45	17.46	18.65	19.92
Total	90.10	101.06	112.04	119.32	127.26	136.42
			C: 1982 projections[c]			
Norfolk	42.24	41.62	45.52	43.31	44.09	45.29
Virginia Beach	24.10	31.49	39.96	41.79	45.24	47.99
Chesapeake	8.72	11.87	14.14	15.56	16.81	17.98
Portsmouth	13.73	14.45	15.06	15.66	16.22	17.00
Total	88.79	99.43	114.68	116.32	122.36	128.26

[a] Letter from Assistant Chief of Engineering Division, U.S. Army, Corps of Engineers, Norfolk District, April 8, 1977, to W. Gardner, Southeastern Virginia District Planning Commission.

[b] City of Virginia Beach, *Water Resource Development Program for Tidewater Virginia*, Prepared for North Carolina-Virginia Water Resources Management Committee (December 14, 1982).

[c] City of Virginia Beach, *Lake Gaston Water Supply Project Environmental Report* (Virginia Beach, 1983).

sector, and (5) distribution losses would be reduced. In fact, all these changes have occurred in the region during the last several years.

Based on these changes, preliminary revisions to the 1977 water use projections were made by the Corps in 1981 (see panel B of table 5-1). Final use projections were made in 1982 (see panel C of table 5-1). The revisions substantially reduced the projections to the rate of growth in population. This is demonstrated by table 5-2, which shows that per capita water use is projected to grow only slightly over the fifty-year period. However, Virginia Beach has expressed concern about the long-term accuracy of the most recent water use projections as follows:

> The demand projections assume public acceptance of long-term water conservation measures. The residents of southeastern Virginia, and more specifically Virginia Beach, have proven that short-term water conservation measures are effective in reducing demand, but have not proven that long-term measures will effectively reduce demands to the level projected by the Corps of Engineers (City of Virginia Beach, 1983).

TABLE 5-2. 1980 and 2030 Projections of Per Capita Water Use (gallons per day)

City	1980	2030
Norfolk	145.5	154.6
Virginia Beach	96.7	97.3
Chesapeake	77.9	82.5
Portsmouth	119.4	131.8

Source: Computed from population and water use projections reported in City of Virginia Beach, Lake Gaston Water Supply Project Environmental Report (Virginia Beach, 1983). (Based on water use projection from table 5-1C.)

Yield Estimates

The yield of a water system that relies on a reservoir is traditionally defined as the constant rate of withdrawal that will just deplete the available reservoir storage by a predetermined percentage, assuming a repetition of the worst drought of record (Linsley and Franzini, 1979). During this drought, withdrawals will exceed inflow and the reservoir will be depleted by the predetermined amount just as the drought ends and inflow begins to exceed withdrawals. Because the record drought has a low probability, a system will produce water in excess of the calculated yield in most years.

Yields for each of the area's water systems in southeastern Virginia were computed during the early 1970s and were recomputed in 1983 by the Corps of Engineers at the request of the Virginia State Water Control Board (Virginia State Water Control Board, 1983). The 1983 estimate by the Corps allowed for a 75 percent depletion of available storage during the record drought of the last fifty years. Columns A and B in table 5-3 show the yield calculations reported for surface-water systems in 1977 and the revised 1983 yield estimates. In addition to surface-water sources, most of the city water systems have groundwater available for use from an artesian aquifer system that underlies the region. Ground water yields, reported in column C of table 5-3, are based on well-pumping capacity. Groundwater has consistently been viewed as a supplemental water source, to be drawn on only if surface-water supplies are depleted to levels unacceptable to water-system managers.[2]

Because streamflows in the watersheds of most of these reservoir systems are unmeasured, estimated surface-water yields have been synthesized through use of flow records from a comparable watershed. As

---

[2]The ability of the regional aquifer system to support sustained groundwater pumping equal to the capacities of these wells is a matter of debate and is discussed in a later section.

TABLE 5-3. Surface-Water Yields and Well Capacity for Southeastern Virginia Water Systems (million gallons per day)

City	Estimated surface-water yields		Well capacity[c] (C)	Prudent management yield	
	1977[a] (A)	1983[b] (B)		Total[d] (D)	Surface water (E)
Norfolk	80.0	66.0	23.5	57.1	36.0
Chesapeake	10.0	10.0	0.0	10.0[f]	10.0
Virginia Beach	0.0	0.0	20.0[e]	0.0	0.0
Portsmouth	27.5	19.0	5.5	18.0	18.0
Total	117.5	95.0	49.0	85.1	64.0

[a] U.S. Army, Corps of Engineers, Norfolk District, Announcement of third public meeting on the water supply study for the south side of Hampton Roads, Virginia, November 21, 1977.

[b] U.S. Army, Corps of Engineers, Norfolk District, "Yield Analysis—Hampton Roads Area" (Norfolk, Va.).

[c] Virginia State Water Control Board, *Historical Review in Current Status of Planning in Southeastern Virginia*, Information Bulletin No. 554 (Richmond, Va., Virginia State Water Control Board).

[d] City of Virginia Beach, *Water Resource Development Program for Tidewater Virginia*, Prepared for North Carolina-Virginia Water Resources Management Committee (December 14, 1982).

[e] Ownership of this well capacity will transfer to city of Suffolk and Isle of Wight and Southampton counties over the next several years.

[f] Use during the 1980/81 drought was 6.8 mgd. This was based upon the maximum demand on the system and not capacity of system to provide water. Because Chesapeake uses a river system, the prudent management yield used for this paper is the Corps' estimate of 10 mgd.

a result, these yield estimates have been viewed with skepticism by local governments. In recent years, the localities have offered an alternative to the standard yield measure, termed the "yield under prudent management."

The prudent management yield for a particular water system is based on average daily production during the 1980/81 drought, the severity of which was similar to the drought of record. Actual production was determined by a drought-management strategy, illustrated here for the Norfolk system, that had the following results:[3]

(1) Water in storage was not drawn down below 50 percent of reservoir capacity. Available storage began to increase late in the drought, indicating that less than 50 percent of storage was used during the drought.
(2) An average of 21.1 million gallons per day (mgd) of groundwater was pumped by Norfolk during the drought.

[3]Memorandum, plus attachments, Sheila Prinderville to Virginia Members—North Carolina-Virginia Water Resources Management Committee, May 2, 1983.

(3) Mandatory and voluntary water demand-reduction programs were implemented in Norfolk and Virginia Beach. Water use reductions of 25 percent, from 76 mgd down to 57.1 mgd, were achieved.

The prudent management yields are shown in table 5-3, column D.

A direct comparison between the Corps' 1983 yield calculations and the prudent management yield estimate is not possible. First, the prudent management yield permits less than 50 percent depletion of reservoir storage capacity, rather than 75 percent. Second, the 1980/81 drought, although nearly as severe, was not the drought of record used in the Corps' calculations.[4] Third, the reported prudent management yield includes groundwater pumping for the Norfolk system. Keeping these differences in mind, groundwater pumping can be subtracted from total prudent management yield to obtain surface-water yields for comparison with the Corps' yield calculations. Surface-water yields during the 1980/81 drought are shown in table 5-3, column E.

These different measures of yield continue to be used. For example, the Virginia State Water Control Board uses the prudent management yield measure to justify its support for the Lake Gaston proposal made by Virginia Beach[5] and uses the Corps' 1983 yield estimates for its public water supply studies for the region (Virginia State Water Control Board, 1983).

The Deficit

Over the past decade, the estimated regional water deficit has varied with changes in use and yield estimates. Throughout the period, the focus has been on surface-water supplies, with groundwater considered as reserve for emergencies. In 1977 the Corps predicted the regional deficit would grow to 66.3 mgd by the year 2030 (183.8 mgd of water use [table 5-1A] less 117.5 mgd yield [table 5-3]). This deficit analysis assumed that surplus capacity at Norfolk and Portsmouth would be available in Virginia Beach and Chesapeake, the cities that were the major source of growth in water use.

After the 1980/81 drought, the localities adopted the lower prudent management surface-water yield estimates and the preliminary revision of the Corps' water use projections. For the year 2030, these commu-

---

[4]According to the Corps' yield analysis (U.S. Army Corp of Engineers, 1983), the 1980/81 drought was the drought-of-record for the Portsmouth water system. For the Norfolk system the drought-of-record occurred in 1954/55. The 1980/81 event was the third-worst drought event of the past fifty years. For the 1980/81 period the Corps computed attainable yields of 73 mgd with a 75 percent reservoir-depletion rule.

[5]Prinderville, *op. cit.*

nities projected a surface-water deficit of 72.4 mgd (136.42 mgd of water use [table 5-1B] less 64 mgd yield [table 5-3]). This deficit analysis still assumed regional sharing of surplus water.

In 1983, the Corps released its final water use projections and surface-water yield estimates. Using these most recent estimates, the 2030 regional deficit would be 33.26 mgd (128.26 mgd of water use [table 5-1C] less 95 mgd of yield [table 5-3]).

Lake Gaston Choice

In 1983, the city of Virginia Beach proposed construction of a 60-mgd pipeline from Lake Gaston in southcentral Virginia (figure 5-1), even though the 2030 projected deficit, incorporating the most recent use projections and yield analysis, is 33.26 mgd. The city's argument for a 60-mgd project was based on rejection of the possibility of sharing any projected surpluses from the Norfolk and Portsmouth systems, as follows:

> Currently there is no guarantee that the entire amount of the surplus will be available during all conditions for use by the City of Virginia Beach and/or the other participating cities and counties. Both Norfolk and Portsmouth are actively pursuing the expansion of their commercial and industrial tax base. Although the southeastern area of Virginia does not actively pursue water intensive industries, it is reasonable to assume that any "surplus" water capacity that may exist would be made available to any new industrial or commercial enterprises specifically wishing to locate in Norfolk or Portsmouth (City of Virginia Beach, 1983).

Accepting the Corps' final water use projections and yield estimates but excluding the regional sharing of surplus water, Virginia Beach argued that it would use 48 mgd by the year 2030. The city also stated that Chesapeake would use 10 mgd from the project and that Franklin and Isle of Wight counties would each take 1 mgd.

The institutional and historical context must be reviewed in order to understand why Virginia Beach projected a 2030 deficit of 60 mgd and proposed Lake Gaston as a water source from among several alternatives.

Institutional Setting for the Choice

Resolution of the problem of urban water supply in southeastern Virginia must be implemented within an institutional framework involving all levels of government. This section summarizes the relevant institutions. A more complete survey is presented elsewhere (Cox and Shabman, 1983).

Role of Local Government. Virginia localities exercise two primary functions relating to water supply: (1) developing and maintaining adequate supplies for their citizens, and (2) controlling water supply development within their boundaries by other political subdivisions.

Local governments in the southeastern area have developed existing water supply facilities and are responsible for future expansion. These localities are authorized by Virginia law to pursue joint solutions to water supply problems on a cooperative basis. However, independent measures have been more common.

A primary institutional factor in interjurisdictional water development is a state law requiring a political subdivision to get the consent of a second subdivision before constructing or operating water supply facilities within the boundaries of the second (Va. Code. Ann. secs. 15.1-37, 15.1-332.1, 15.1-875, 15.1-1250.1). This provision is a factor in determining the feasibility of any proposal to transfer water into the cities of southeastern Virginia.

These consent requirements do not provide an absolute veto power. Where consent is denied, the requesting jurisdiction has the right of appeal to a special panel made up of three Virginia circuit court judges. The special court may grant consent for the proposed action under what it deems fair and reasonable terms and conditions. However, the panel is not empowered to adjudicate individual water rights claims that must be resolved in the courts according to state law (to be discussed below).

The State Role. State water management institutions view urban water planning and management as primarily a local responsibility. The State Water Control Board, Virginia's administrative water resources agency, has conducted water resources planning since 1972, but its emphasis has been on water quality. As a result, the basic information necessary for developing urban water plans is incomplete. For example, the yields of the aquifer systems underlying the eastern regions of Virginia remain unclear. Recent legislation has mandated increased attention to urban water planning (Va. Code. Ann. sec. 62.1-44.38) but this legislation will not immediately remedy the information deficiencies.

A basic component of the institutional framework is the state's water allocation law. This body of law remains primarily common law, where water rights exist as an element of land ownership. Conflicting claims to water are adjudicated by the courts. An exception to the common law approach is created by the Virginia Groundwater Act of 1973 (Va.

Code. Ann. sec. 62.1-44.83 *et seq.*). The act requires a state permit for groundwater withdrawals within designated areas, which now include Virginia's Eastern Shore and southeastern Virginia. However, a state attorney general's interpretation of the groundwater act, an interpretation perhaps unintended by the act's authors, has ruled that groundwater withdrawals for public supply are covered by the act's exemption for domestic use and therefore are outside state permit authority.

Water rights of public suppliers located within management areas apparently remain subject to the state's common law. Neither the groundwater act nor the courts have addressed the rights of groundwater permit holders vis-à-vis common law rights. This gives rise to uncertainty about the security of both groundwater permit rights and common law rights.

Adding to this uncertainty is the fact that common law groundwater rights have not been well defined by the courts. At this time, it appears that the Virginia Supreme Court favors the reasonable-use groundwater doctrine (*Couch* v. *Clinchfield Coal Corp.*, 139 S.E. [1927]). This doctrine essentially allows any traditional use of groundwater on the land from which it is pumped, without regard for impact on the availability of groundwater for others or on surface-water users should a hydraulic interconnection exist. However, the doctrine generally prohibits export of water for offsite use if other users of the aquifer are adversely affected. This export restriction subjects groundwater use by public water systems to legal challenges because the water withdrawal site is generally not the use site.

Water from Virginia's streams is controlled by the common law riparian doctrine, in which water rights arise as a consequence of ownership of land in contact with a body of surface water. The riparian right is not quantified but permits all use of water on riparian land considered "reasonable." Reasonableness is a relative concept requiring sharing water among the riparian landowners. Reasonableness must be determined on a case-by-case basis through lawsuits initiated by parties who argue that their riparian rights have been adversely affected by another party's water use (*Virginia Hot Springs Co.* v. *Hoover*, 130 S.E. 408 [1925]).

In theory, the riparian doctrine can constrain public suppliers. First, public water supply is not viewed as a reasonable water use (*Town of Purcellville* v. *Potts*, 19 S.E. 2d 700 [1942]) and, therefore, is not lawful if it causes injury to water use recognized as valid under the doctrine. Second, the right to use water on riparian land may limit transfer of water to a water service area. However, neither restriction is operative if there is no injury to holders of valid riparian rights (*Town of Gordonsville* v. *Zinn*, 106 S.E. 508 [1921]). Determination of injury to another water user in a particular case must be established in court proceedings initiated by the party claiming injury. Most of the public suppliers in Virginia operate within this limitation to the doctrine's restrictions on water use. In addition, public suppliers can acquire water rights, if necessary, through eminent domain even though public use has been determined to cause injury to other water users (*Town of Purcellville v. Potts, supra*).

Interstate Institutional Mechanisms. Most of the water supply sources for the southeastern area of Virginia are from rivers flowing into North Carolina. In addition, the aquifer underlying southeastern Virginia extends into northeastern North Carolina. Virginia and North Carolina have not created formal compacts for water management, but the states do have an advisory committee to review mutual water management concerns. The committee's most recent organizational structure was established in 1982 and consists of the twenty-six-member North Carolina-Virginia Bi-State Water Resources Management Committee. Interstate liaison committees examined groundwater use and the use of water in the Roanoke River basin early in 1983.

The Federal Role. Federal government responsibilities affecting urban water supply development include regulating construction in navigable waters and the resolution of interstate water supply conflicts within the federal courts. In addition, Congress may enact specific authorization for a federal agency to assist in the development of detailed plans for urban water supply management.

Federal regulation of water project construction is based on a number of statutes and administered by several agencies. A primary form of control is a permit requirement established under section 404 of the Clean Water Act (33 U.S.C.A. 1251 *et seq.*). The main focus of 404 permit review is assessment of a project's dredge-and-fill activity on water quality. Further, the applicant's project must also comply with the National Environmental Policy Act (42 U.S.C.A. 4321 *et seq.*), which may dictate preparation of an environmental impact statement if significant environmental effects are anticipated. Other constraints on permit issuance include the need to consider the impact on fish and wildlife (16 U.S.C.A. 661 *et seq.*), protection for the habitat of a designated endangered or threatened species (16 U.S.C.A. 1531 *et seq.*), the impact of a project on a stream segment included or being considered for inclusion in the Wild and Scenic Rivers System (16 U.S.C.A. 1271 *et seq.*), and the impact on sites included in the National Register of Historic Places (16 U.S.C.A. 470 *et seq.*). These examples indicate the broad scope of considerations encompassed by federal regulatory procedures.

Because no formal compact exists between Virginia and North Carolina, the federal courts provide the primary forum for resolution of conflicts over legal rights in interstate waters. Individuals adversely affected by water use activities in another state may, under certain conditions, bring suit in federal district court (28 U.S.C.A. 1 *et seq.*). Also, suits between states themselves are possible within the original jurisdiction of the U.S. Supreme Court (U.S. Constitution Art. III, sec. 2). The Court has resolved several lawsuits involving water use conflicts between states by application of its concept of "equitable apportionment." This concept requires a determination by the Court of a fair division of interstate waters. (Examples include: *Connecticut* v. *Massachusetts*, 282 U.S. 660 [1931] and *New Jersey* v. *New York*, 256 U.S. 296 [1931].)

Congress has authorized the U.S. Army Corps of Engineers to conduct studies of urban water supply and related issues in several areas of Virginia. For studies on urban water supply, the Corps estimates future water use and current water system capacity and screens alternatives for addressing projected deficits. These analyses can provide a supplement to local or state studies, or parts of Corps studies can replace studies that might otherwise be conducted by local or state agencies. However, the Corps' analyses will follow federal water planning guidelines even though no federal project is anticipated. Thus, local and state agencies have only limited influence on the study procedures and conclusions. On the other hand, Corps studies do provide technical analyses that are perceived to be unbiased, especially when no federal project is at stake.

The Search for Alternatives

During the past several years, the search for alternatives to meet a projected deficit has focused on two interrelated issues: (1) the use of groundwater and (2) the relative supply of water from within and outside the region. However, the changing deficit estimate, the institutional environment, and a series of droughts interacted to affect the consideration of alternatives and the ultimate selection of the Lake Gaston pipeline.

Groundwater. Because the potential for using groundwater within the region remains unclear at present, the area governments prefer surface-water alternatives. This preference developed after a decade in which the state failed to adopt a clear policy on groundwater use, failed to clarify legal rights to groundwater, and failed to provide a clear description of the aquifer's capacity and how it would be affected by use. Without state initiative, conflicts between local governments over groundwater have escalated.

The unregulated use of groundwater in the early 1970s stimulated passage of the Virginia Groundwater Act. In the process of implementing the act, the Water Control Board developed a groundwater model for predicting the impact of additional pumping on artesian pressures and existing wells within southeastern Virginia. The board's analyses suggested that significant increases in pumping would adversely affect existing wells and ultimately pose a threat to the hydrologic characteristics of the aquifer system itself. By the mid-1970s, the Water Control Board lost confidence in the existing model but continued to hold the position that additional groundwater pumping would have adverse consequences. Discomfort with the Water Control Board position prompted a legislative water study commission to contract with a private consultant, Geherity and Miller (1979), to determine maximum acceptable withdrawals from the aquifer. The results of this 1979 study opposed the Water Control Board's position and stated that as much as an additional 100 mgd could be withdrawn without unacceptable detrimental effects. However, careful spacing of new wells would be necessary.

During the 1980/81 drought, conflict between Norfolk and the city of Suffolk arose over Norfolk's proposal to construct wells on land owned by Norfolk but lying within the city of Suffolk. Suffolk denied Norfolk a conditional-use permit. Norfolk then proposed constructing wells on land owned by the U.S. Navy, where Norfolk contended Suffolk's land-use controls would be inapplicable. Thereupon, Suffolk brought suit to enjoin Norfolk from drilling wells on the Navy's property. The basis for Suffolk's motion included allegations that Norfolk's actions violated Suffolk's zoning ordinances and common-law groundwater rights. Suffolk also brought suit against the Secretary of the Navy, alleging that Navy authorization for Norfolk's wells on federal property, without an environmental impact statement, violated provisions of the National Environmental Policy Act. In turn, Norfolk brought suit to enjoin Suffolk's opposition to its wells. This suit was based on several grounds, including allegations that property owned by Norfolk was being taken and that Suffolk was attempting to monopolize trade and commerce (Hrezo, 1981).

The long drought forced Norfolk and Virginia Beach, in desperation, to seek an out-of-court settlement on terms dictated by Suffolk. In addition, Isle of Wight and Southampton counties were asked to allow well development in their jurisdictions. The result was construction of

20 mgd of well capacity in the three western localities, with the proviso that the wells could be pumped only during a drought emergency. Also, the host jurisdictions would receive payments based on the amount of water pumped. Finally, ownership of the wells, paid for by Virginia Beach, would transfer to the host jurisdictions by the year 2001. After this time, the host jurisdiction would determine Virginia Beach's access to groundwater.

In the midst of this dispute, Governor John Dalton's office, in the absence of direct involvement by the State Water Control Board, proposed three general policy guidelines for interjurisdictional negotiation. The guidelines provided for the jurisdiction receiving new water supplies to pay for (1) land and physical structures necessary to secure water, (2) compensation to jurisdictions supplying water, and (3) costs of correcting damage to wells adversely affected by increased pumping (Virginia State Water Control Board, 1981).

The governor's proposed guidelines, in conjunction with the findings of the Geherity and Miller study (1979), were incorporated into the only Water Control Board water supply proposal made for the area. This proposal suggested development of 42 mgd of groundwater to be followed by development of surface-water supplies. In its final report, the board made the following statement about yield of the region's groundwater:

> It seems reasonable, therefore, to expect that *if* with proper well spacing and capacity the additional 50 to 100 mgd projected by the Geherity and Miller, Inc. report can be safely developed . . ., development of the 42 mgd required in this plan would leave ample water for the host water resource jurisdictions to meet their long term needs. (Emphasis added.) (Virginia State Water Control Board, 1981.)

However, after the bitterness engendered by the previous groundwater-use conflict, local governments gave little consideration to this proposal, and the Water Control Board did not seek its implementation. Moreover, the board never explicitly endorsed the Geherity and Miller study, leaving some doubt about the agency's position on the feasibility of additional groundwater development.

During 1980/81, the states of Virginia and North Carolina initiated a joint investigation of groundwater availability that involved development and use of an analog groundwater model. The model was calibrated in 1983. Based on the results of the analog model, the groundwater technical committee of the North Carolina-Virginia Bi-State Committee recommended stabilization of the current conditions in the aquifer. It suggested ". . . that groundwater should not be used as part of future additions to the base system for major water supply systems in southeastern Virginia. . .,"⁶ although the committee felt intermittent pumping of groundwater would be acceptable to supplement surface-water sources.

However, the committee felt that its conclusions should be treated as tentative until a more sophisticated digital groundwater model was developed and state policy on the use of groundwater was clarified, which might permit further reductions in artesian pressures and attendant drawdown of water levels in wells. The committee's discussion of this was summarized as follows:

> There was a great deal of discussion centering around whether or not stabilization or the cessation of drawdown should occur at present levels or whether we should recognize that additional demands on the system could be supported . . . . Discussion also included the possibility of establishing a water level, above the top of the confining layer, below which water levels would not be allowed to fall. It was recognized by the committee in general that this amount of "free board" above the top of the confining layer could be established somewhat arbitrarily and refined through work with a digital model. In essence the discussion centered around establishing a philosophy for groundwater management on an areawide basis (Virginia State Water Control Board, 1981).

With this bistate report in hand, the State Water Control Board endorsed the proposal of Virginia Beach to ignore groundwater as a supply source and to tap the surface waters of Lake Gaston. Nonetheless, over 45 mgd of groundwater well capacity has now been developed as part of the Norfolk water system and is a potential water source during droughts. This capacity could also be used on a continuing basis, although it would be accompanied by a reduction in artesian pressures.

Water Importation into the Region. In the early 1970s, proposals to expand the region's water supply typically recommended extension of surface-water supplies within the region. An example was construction of a reservoir on the Blackwater River north of Franklin (Whitman, Requardt, and Associates, 1971). Application for necessary federal regulatory approval was made in 1974. However, State Water Control Board concern about environmental impact of this project resulted in its abandonment.

Another consultant's study published in 1975 (Henningson and coauthors, 1975) analyzed a wide range of alternatives. Options such as

---

⁶Memorandum, R. A. Masiello to J. S. Cragwell, Jr., April 12, 1983, Technical Liaison Groundwater Committee Meeting.

desalinization, wastewater reuse, and groundwater development were rejected because of high cost, technical or political infeasibility, or uncertainty about the reliability of the source. More-detailed analysis focused on surface-water sources, including withdrawals from outside the region in the Roanoke and James river basins and withdrawals from within the region from the Chowan (Blackwater and Nottoway) and Northwest river basins. The study recommended a 10-mgd withdrawal from the Northwest River and a similar withdrawal from Lake Gaston in central Virginia.

The Northwest River withdrawal plan was implemented by the city of Chesapeake, but the Lake Gaston alternative was not. Instead, area governments decided to wait for the recommendation of the Corps of Engineers' study that had been authorized in 1974. However, the Corps' study was delayed by the modifications to federal water planning procedures discussed earlier.

The drought of 1977 made the city of Virginia Beach especially sensitive to its vulnerability to regional inaction. As a result, Virginia Beach contracted for independent studies of alternatives for providing the city with its own supply. After these studies were completed, the 1980/81 drought occurred, and Norfolk made significant reductions in its delivery to Virginia Beach. These reductions heightened Virginia Beach's concern about its dependence on the Norfolk system. The regional conflict over drilling emergency wells also occurred during this period.

One result of the 1980/81 drought, then, was Virginia Beach's increased distrust of its neighboring jurisdictions as water suppliers. In September 1981 the city issued a position paper suggesting that it would purchase treated water from the Appomattox River Water Authority, bypassing the Norfolk system and abandoning its wells in Suffolk (City of Virginia Beach, 1981). Virginia Beach defended this position, stating that Norfolk was "unwilling to guarantee a baseline supply" (City of Virginia Beach, 1981). The implications of the proposed Virginia Beach action were significant for Norfolk. Norfolk's treatment plant has the capacity to treat 108 mgd of water. Without Virginia Beach as a customer, only a fraction of that capacity would be used, resulting in financial loss for Norfolk. This Virginia Beach position paper encouraged Norfolk and Virginia Beach to reconcile their differences and jointly support a proposed 60-mgd pipeline from Lake Gaston. In 1983 Virginia Beach filed a request for a section 404 permit for pipeline-related construction with the Corps of Engineers.

Lake Gaston, a privately owned power reservoir, is one of a series of impoundments on the Roanoke River along the Virginia-North Carolina border (figure 5-1). The Corps' operation of the upstream Kerr Lake maintains lake levels at Gaston for hydroelectric power, flood control, and environmental quality. The lakes are also used extensively for recreation.

However, proposals for withdrawal from Lake Gaston have not been without opposition. Throughout the years that Lake Gaston was considered as a water source, the citizens in the Roanoke River basin actively opposed the withdrawal. The basis for their opposition was concern that a water transfer would result in substantial drawdowns at Kerr Lake, reducing property values on the lake shore, damaging recreational fishing in the lake, and harming fish spawning downstream of the impoundments. The damage to recreation and fish habitat was of particular concern because the surrounding counties' economies are almost wholly dependent on agriculture and the tourism attracted by the impoundments. These concerns have been voiced for nearly a decade, but until recently no credible analysis of their validity was made by the states or by the Corps.

As part of the Bi-State Advisory Committee activities, these issues were addressed in 1983 by the state of North Carolina in cooperation with the city of Virginia Beach, the Virginia Water Control Board, and the Corps of Engineers' Wilmington Office. In general, the findings are that a 60-mgd withdrawal will reduce Kerr Lake levels by 0.3 feet in an average year and 1.2 feet during the drought of record. In comparison, operation of the reservoir for flood control can result in lake-level fluctuations of as much as 22 feet, and normal operations will result in changes of 4–5 feet in lake levels during the recreation season. The 60-mgd transfer would represent only 1.2 percent of the river's average annual flow. In a worst-case scenario developed by North Carolina based on a drought of record and particular operating rules, streamflow below Lake Gaston would be reduced by 8.5 percent.[7] Reviewing these effects, the Corps issued a section 404 permit for the pipeline in January 1984, without objection from the U.S. Fish and Wildlife Service or the U.S. Environmental Protection Agency. The Corps found no significant adverse environmental effects and did not require an environmental impact statement. However, North Carolina has initiated court action to require filing of a statement.

In addition to challenging the issuance of a 404 permit, pipeline opponents may initiate lawsuits. Actions under riparian water law, to be successful, would have to demonstrate that the pipeline diversion will harm riparian landowners. This argument is unlikely to be persuasive because of the lack of significant impact on lake levels and streamflows.

[7]Letter, J. W. Grimsley, Secretary, North Carolina Department of Natural Resources and Community Development to Colonel Ronald E. Hudson, Norfolk District, U.S. Army Corps of Engineers, September 2, 1983.

Potential challenges by North Carolina in federal court on the question of that state's interests in the basin's waters may be broader in scope than private lawsuits. Based on the results of assessments done for the 404 permit review process, however, these challenges would be unlikely to result in prohibition of the transfer.

The Lake Gaston Pipeline Versus Local Supply: A Cost Comparison

To prevent water-use conflict, investment in water system development may be greater than necessary. When this occurs, these added costs can be attributed, as an economic efficiency cost, to existing institutional obstacles to conflict resolution. In considering whether this has occurred in southeastern Virginia, it is first necessary to document that in a different institutional setting, a less costly alternative than the Lake Gaston pipeline could have been chosen. This investigation involves three steps. First, a comparison is made between the manner in which the projected deficit would be closed by the Lake Gaston pipeline and the alternative. Second, the increased costs of the pipeline over the most likely alternative are discussed. Then the argument is presented that additional costs are being incurred in order to avoid water-use conflict.

A Conjunctive Water Use Alternative

Whatever development strategy is adopted, Virginia Beach insists that it close the projected water deficit. Regional annual surface-water deficits computed at 10-year intervals are reported in table 5-4. The projected surface-water deficit under a recurrence of the drought of record,

TABLE 5-4. Surface-Water Deficit for Southeastern Virginia, 1980 to 2030[a] (million gallons per day)

Year	75% reservoir depletion[b]	Less than 50% reservoir depletion[c]
1980	+6.21	−24.79
1990	−4.43	−35.43
2000	−19.68	−50.68
2010	−21.32	−52.32
2020	−27.36	−58.36
2030	−33.26	−64.26

[a] Computed as sum of 1983 final water use projections (table 5-1C) less surface-water yields (table 5-3).
[b] Basis for U.S. Army Corps of Engineers yield estimate.
[c] Basis for prudent management yield estimate.

TABLE 5-5. Water Surplus or Deficit After Implementation of Surface-Water Development Alternatives (million gallons per day)

Year	Lake Gaston pipeline (60 mgd)		Conjunctive use alternative (30 mgd)[b]
	75% reservoir depletion	Less than 50% reservoir depletion[a]	
1980	+66.21	+35.21	+36.21
1990	+55.57	+24.57	+25.57
2000	+40.32	+9.32	+10.32
2010	+38.68	+7.68	+8.68
2020	+32.64	−1.64	+2.64
2030	+26.74	−4.26	−3.26

[a] Computed as deficit shown in table 5-4 plus 60 mgd of added yield.
[b] Computed as deficit with 75% reservoir depletion (from table 5-4) plus 30 mgd of added yield.

based on depletion of 75 percent of reservoir storage, rises from 4.43 mgd in 1990 to 33.26 mgd in 2030. Under the prudent management approach (where less than 50 percent of reservoir storage was used during the 1980/81 drought), the surface-water deficit rises from 24.79 mgd in 1980 to 64.26 mgd in 2030. Also, Norfolk, Virginia Beach, and Portsmouth currently have well capacity of 49 mgd, of which 20 mgd will eventually revert to other jurisdictions.

Construction of the Lake Gaston pipeline would ensure a surface-water yield increase of 60 mgd. The effect of the added 60 mgd on the deficit is shown in table 5-5. Clearly, the pipeline provides adequate surface water to meet the projected deficit.

Several alternatives to the pipeline have been considered during the last decade. For this analysis, an alternative is developed that is a variation on the Water Control Board's 1981 groundwater plan combined with early proposals to further develop the Blackwater River.

Termed the conjunctive use alternative, this plan would use the existing 49 mgd of groundwater capacity as a supplemental source, interconnect the Portsmouth and Norfolk water systems, develop a drought management plan, and expand current use of surface water from the Blackwater-Nottoway river system. The Portsmouth and Norfolk systems can be connected by a 15-mile pipeline (Anderson, 1978), which would fully integrate the area's water sources. In effect, a regional water system would be established.[8]

[8] In fact, interconnection and coordinated management of the various systems could result in yields for the regional system that would exceed the arithmetic sum of current yield estimates. This was found to be the case in the Washington, D.C., metropolitan area (Sheer, n.d.). However, this possibility has not been examined for southeastern Virginia and, at present, cannot be considered in the estimation of yields from the conjunctive use alternative.

The drought-management plan would be based on a systematic evaluation of the region's water systems and would be a strategy for joint management of reservoir storage, river pumping, and well capacity. The plan would assume that water system managers would be willing to use more than 50 percent of available reservoir storage. This approach would reduce the need for water use restrictions as severe as those required in 1980/81.

An analytically based drought-management plan would draw on optimization and simulation techniques developed before a drought for expanding joint yields of the separate reservoir, river, and well systems. The system would include a procedure for risk analysis, based on hydrologic records and demand expectations, to identify the start of potential droughts and quantify the risks of continued drought. With this information, day-to-day decisions on necessary use restrictions and system operation would be made on a sound analytical base. (For discussion of the successful use of such an approach, see Sheer, n.d.)

For years, further development of the Blackwater and Nottoway rivers has been considered and rejected. A prior concern of the State Water Control Board has been that increased diversion of 60 mgd would reduce flows for assimilation of industrial waste. However, a recent study shows that skimming high flows to store for later use will not affect downstream capacity to assimilate waste (Sheer and Erlich, 1982). In 1981, the executive director of the board expressed no opposition to a smaller withdrawal of 30–40 mgd from the river system (Davis, 1981). Thus, the river could provide an additional 30 mgd of water. For purposes of this analysis, 11 mgd of added yield would come from expanding the pump capacity on the Blackwater and Nottoway rivers. This increase was supported by the 1983 Corps of Engineers' yield study. The rest of the 30 mgd would come from construction of a small 5-billion-gallon reservoir. Previous studies have indicated that this would add 19 mgd (Whitman, Requardt, and Associates, 1971).[9]

Table 5-5 shows the effect of this 30-mgd increase on the surface-water deficit during a recurrence of the drought of record. By the year 2030, a drought would require some use of groundwater or water-use

---

[9]This 19-mgd yield estimate, made in 1971, was based on the drought that ended in 1966, which for the Blackwater-Nottoway drainage area was "slightly worse" (Whitman, Requardt, and Associates, 1971) than the drought ending in 1955. The 1955 drought was used by the Corps for computing drought-of-record yields for the entire Norfolk system. Therefore, the 1971 yield analysis for a 5-bg reservoir is based on a drought event comparable with the 1983 yield analysis for the existing system.

The 1971 study, however, used a 65 percent reservoir-depletion rule for the system in computing the 19-mgd yield. Thus, this yield analysis, like the Corps', concludes that more than 50 percent of reservoir capacity would be drawn upon in a drought.

reductions or both. However, with a drought-management plan in effect, groundwater would probably be drawn upon before available reservoir storage was fully depleted. Therefore, in years prior to 2030, some groundwater would be used in conjunction with surface water, even with a drought less severe than the drought of record. However, the existing 49 mgd of well capacity would prove adequate.

As this discussion indicates, either the conjunctive use alternative or the Lake Gaston pipeline would provide a reliable source of water to meet the current projection of the regional deficit. As will be argued later, reluctance to select an alternative to the Lake Gaston pipeline can reflect a desire to avoid water-use conflicts rather than a lack of confidence in the alternative solutions.

Estimated Cost Differences

In the Lake Gaston pipeline plan, existing groundwater capacity is abandoned and a substantial portion of existing reservoir storage remains unused; in contrast, the conjunctive use alternative makes use of the existing capital investment. Because the pipeline to Lake Gaston duplicates existing system capacity, it can be expected to cost more than the conjunctive use alternative. Precise estimates of the costs of the two alternatives are not available. However, sufficient information does exist to approximate the difference.

Operating Costs. This cost comparison will be based mainly on construction costs. Available planning documents report construction cost as one basis for screening the solutions, and annual operating information is thus limited. Nonetheless, it is possible to draw some general conclusions about the operating costs of the two schemes. First, differences in the quality of raw water that they deliver will result in higher water treatment costs for the conjunctive use alternative. Blending of ground and surface water during the infrequent drought periods when groundwater will be used can raise the operating costs for treatment plants. Further, the expansion of current water use from the Blackwater River can require special treatment for concentrations of dissolved organics. But because this drainage is already a water source for the area, only an extension of current practice would be needed.

Second, there are operating costs unique to each alternative. Use of the Lake Gaston pipeline will result in charges for forgone hydroelectric power and for increased reservoir operating costs. At an average flow rate of 40 mgd, annual costs of $1,103,840 have been estimated (Maguire, 1982). However, there is no basis for estimating the time pattern of pipeline use, so this potentially large cost has not been included in the comparative cost analysis. On the other hand, the cost for continually operating a drought-management plan was not included in the annual cost of the conjunctive use alternative.

Third, groundwater pumping costs will be incurred only during random drought emergencies. Therefore, the expected value of these costs will be a small proportion of the total cost of the conjunctive use alternative. A large component of operating costs for both plans would be for pumping surface water. However, there is little reason to expect substantial pumping cost differences between alternatives. Therefore, omitting these annual costs from the analyses should have no substantive effect on the results.

Construction Costs. In 1982, construction costs for the raw water transmission system from Lake Gaston to an existing transmission line in Suffolk were estimated to be $124.5 million (City of Virginia Beach, 1982a). This cost includes $7.8 million for two pump stations; $91.7 million for the land and pipeline; and overhead costs (contingency, legal, administrative, and engineering) of $25 million, calculated as 25 percent of construction cost.

Construction costs for the conjunctive use alternative would include expenditures for a storage facility, pumps, a pipeline capable of carrying 30 mgd of surface water to the existing transmission line at Suffolk, and a pipeline connecting the Portsmouth and Norfolk systems. Wells and transmission lines from the wells are already in place. Construction costs for the 5-billion-gallon reservoir to provide 19 mgd cannot be precisely made without field surveys and preliminary engineering studies. However, in 1971 a consultant to Norfolk "examined the cost of several other reservoir projects and . . . [found] the present-day cost is approximately $1 million per 1 bg (billion gallons) of storage" (Whitman, Requardt, and Associates, 1971). In 1971 dollars, the proposed reservoir was estimated at $5 million. A construction cost index was used to inflate this cost to 1982 dollars for comparison with the Lake Gaston pipeline costs, making the reservoir cost $14.65 million.[10]

A 30-mgd pipeline and the necessary pumps constitute the other significant costs of the conjunctive use alternative. Pump and pipeline costs will vary with pump size, pipeline size, and pipeline length. The pumps and pipeline needed for the conjunctive use alternative would be smaller than those used for the Lake Gaston transfer, and substantially less pipeline length is needed.

---

[10]Comparison of the construction cost indices for 1971 and 1982 for Baltimore, Maryland, indicated construction costs in the region rose by 293 percent during the period (Engineering News Record, 1971 and 1982).

Because of the short distance from the river to the existing water line at Suffolk, it is assumed that no intermediate booster pump would be needed. Therefore, approximate costs for a single 30-mgd pump station will be used here. In 1982, pump cost estimates for the Lake Gaston pipeline were approximately $60,000 per mgd. For purposes of this discussion, pump costs for the conjunctive use alternative in 1982 dollars, will be set at $1.8 million (30 mgd × $60,000 per mgd).

No estimates for a 30-mgd pipeline or a pipeline connecting Portsmouth to Norfolk are available. For this comparison, the per-mile cost of the 60-mgd Lake Gaston pipeline, from its intersection with the Blackwater River to the existing line at Suffolk, will be used as a crude approximation of the pipeline costs for the conjunctive use alternative. The distance from the Blackwater River to existing lines is approximately 15 miles, or 17.5 percent of the 85-mile pipeline length. This 17.5 percent of the $99.5 million total cost for the pipeline in the Lake Gaston transfer yields an approximate pipeline cost of $17.5 million for the conjunctive use alternative. The $17.5 million is also used to approximate the cost for a 15-mile line to connect the Norfolk and Portsmouth water systems. The total construction cost for the reservoir and pipelines in the conjunctive use alternative is $51.3 million. Applying the same 25 percent overhead factor as was used in the Lake Gaston pipeline estimate makes the final cost approximately $64.1 million.

Although the cost estimates used to compute these values must be treated as approximations, the results suggest that substantial efficiency losses will arise in southeastern Virginia because of institutional barriers that led to selection of the Lake Gaston alternative. The increase in capital cost over the conjunctive use alternative is around $60 million. It is not likely that more precise figures would alter the general conclusion that substantial aggregate efficiency losses have been imposed by southeastern Virginia's water management institutions.

From another perspective, the $60 million increase in capital investment has not threatened political support for the proposal within the region. One reason for this is the modest effect on customers' water bills from increased annual debt service costs to pay for the pipeline. Using a 25-year payback period and 9 percent interest, annual debt service is increased by $6.11 million by the choice of the Lake Gaston pipeline over the conjunctive use alternative. In 1990, annual water use of 11,494 million gallons is projected for the city of Virginia Beach (365 days × 31.49 mgd). If the added annual cost of $6.11 million is recovered by raising water rates, the increase in the rate structure would be $0.532 per 1,000 gallons in 1990 ($6.11 million dollars divided by 11,494 million gallons). For example, a household in Virginia Beach using 5,000 gallons per month would pay an additional $31.92 per year for its water cost.

Although this number is not insignificant, it does indicate that the large aggregate efficiency losses are not meaningful for individual customers.

Criteria for Selecting the Lake Gaston Pipeline

Several years of conflict over expanding the urban water systems in southeastern Virginia dictated the criteria used for selecting the Lake Gaston pipeline. These criteria favored alternatives that promised to minimize jurisdictional conflict. Although interdependent, the criteria were distinguishable. Virginia Beach sought to choose the alternatives that (1) minimized the use of groundwater, (2) minimized the need for regional drought management, (3) minimized the likelihood of successful legal or administrative challenges, and (4) minimized the effects of water-use projection errors. The opportunity to spread significant aggregate cost differences over a large population base reduced the importance of cost as a criterion for choice. Once alternatives with clearly unacceptable costs (such as desalinization) were removed from consideration, relative costs became unimportant in the selection process. Based on the history of the Lake Gaston selection, this was clearly a defensible position.

Minimizing the Use of Groundwater. Conflicts over groundwater as a source of supply have not been resolved under existing institutions. The state's groundwater law includes no clear policy governing either acceptable reductions in artesian pressures or the possibility of a new water user compensating users of existing wells damaged by reduced artesian pressures. In fact, no generally accepted analyses exist of the likely effect of increased pumping on wells.

Meanwhile, existing groundwater law and the absence of state water policy leadership perpetuates conflict among political jurisdictions and groundwater users. As a result of agreements reached during the 1980/81 drought, future use of groundwater in the western counties will require payments by the eastern localities under terms still to be negotiated. But the only mechanism that now exists to support negotiation between parties is a provision in Virginia law for the jurisdiction to appeal denial of a proposed transfer of water to a special panel of judges. However, even agreements reached in this forum would not secure the water for Virginia Beach. Under existing groundwater common law, private parties, not localities, hold ownership rights, and private lawsuits may still be filed in the courts.

The Virginia Groundwater Act could have been a means for state intervention in the groundwater disputes that preceded the decision to choose the Lake Gaston alternative. However, municipal use is exempted from the act's permit authority. This situation was described by Virginia Beach as follows:

> The transport of groundwater across political subdivision boundaries has proved to be at least as, if not more, volatile an issue as the transport of surface water. There is no federal permit process for groundwater withdrawal and current Virginia law does not clearly address disputes that arise between competing interests for groundwater. Strictly interpreted, Virginia State law would imply that a land owner should be able to withdraw as much groundwater as he is physically able to use and dispose of. However, during the 1980–81 drought, Norfolk attempted to drill wells on land it owned in the rural jurisdictions of southeast Virginia. The localities objected stating that they wanted to own and control the wells as well as receive payment for water withdrawn. These conditions proved to be unacceptable to Norfolk and a lengthy and expensive court action ensued.
>
> Faced with a possible 50% water allocation for the summer of 1981, Virginia Beach agreed to the conditions set down by the rural localities and five emergency wells were drilled. The contracts for the operation of the wells (which expire within periods of 5 to 10 years) are not suitable for long-term water supply agreements. Institutional and legal problems similar to those encountered during the 1980–81 drought should be expected if the emergency wells are to become a permanent part of the water supply (City of Virginia Beach, 1983).

Therefore, Virginia Beach sought alternatives that would minimize the need to draw upon the 49 mgd of well capacity now in place in the region. Particular emphasis was placed on avoiding use of the 20 mgd of that capacity built during the 1980/81 drought.

With the pipeline, use of groundwater is virtually eliminated, even with a recurrence of the drought of record. In contrast, the conjunctive use alternative relies on groundwater as a component of a drought management strategy. In fact, a drought management plan would not be needed if the pipeline is constructed because, even with use of less than 50 percent of reservoir storage, adequate water would be provided by the pipeline. For a jurisdiction that wants to avoid the use of groundwater, the pipeline alternative is superior.

Minimizing the Need for Drought Management. During the 1980/81 drought, Virginia Beach and Norfolk water system managers had to make a series of water use restriction and reservoir operation decisions without benefit of a previous drought emergency plan. Lacking a preexisting plan, management had to proceed with limited information. With hindsight, it is clear that, in 1980/81, less than 50 percent of the available reservoir storage was used rather than the 75 percent used in the yield analyses, and that the full amount of groundwater capacity was not tapped. However, since 1980/81, no local effort has been made to develop a drought-management plan that would increase reservoir use. More important, current institutional arrangements provide little incentive for developing a regional drought-management strategy. Therefore, if a substantial drought recurred, Virginia Beach could expect that again full use would not be made of computed system yields, and that it could expect either to force a substantial cut in water use or to use groundwater.

Minimizing Successful Challenge. Challenges to implementation of a water supply alternative may be brought within administrative or court proceedings. In the current institutional setting, two primary bases exist for such challenges. First, a project opponent might document occurrence of socially unacceptable environmental impacts under the administration of section 404 of the Clean Water Act. The second basis for challenge is water law, under which opponents can document that the proposal's implementation would interfere with their water rights. But court and administrative review proceedings can be of uncertain duration, with unpredictable results.

Because of its urgency to develop more water system capacity, Virginia Beach sought an alternative that would minimize the success of any environmental or water law challenges. The Lake Gaston pipeline meets this goal. At this time, the issuance of a section 404 permit is being challenged in court. However, there is little probability of ultimate success for this challenge, given the lack of objection to the pipeline by the U.S. Environmental Protection Agency and the U.S. Fish and Wildlife Service. Another possible challenge to the pipeline could be made under riparian law. The success of that challenge will require riparian owners to document damage. Further, North Carolina might challenge the transfer on grounds that it results in an inequitable division of the basin's waters between the two states. Neither of these legal challenges would appear to have much chance for success, given the small magnitude of the proposed diversion in relation to the river's flow.

In contrast, the conjunctive use alternative is likely to perpetuate the historical conflicts. Resolution of groundwater conflicts would be necessary, an unlikely event given the lessons of recent history. Moreover, additional surface-water development within the region would require a new round of political negotiations with neighboring governments, raise the potential for riparian challenges, and increase exposure to environmental opposition, especially if construction of a new impoundment is necessary.

Minimizing the Effect of Projection Error. Virginia Beach remains skeptical of the demand-reduction assumptions in the water use projections made by the Corps of Engineers. In Corps water use and surface-water yield estimates, the maximum regional deficit during the drought of record was projected to be 33.26 mgd in 2030; therefore, construction of a 60-mgd pipeline would result in significant excess capacity even in the year 2030. Indeed, excess capacity is greater than that shown in table 5-5, because surface water yields will exceed those calculated for the drought of record in almost all years.

For the conjunctive use alternative, excess capacity declines from 36.21 mgd in 1980 and disappears by 2030. Of course, this situation would only occur during a repeat of the drought of record; in most years the margin against projection error is greater. However, if Corps projections are in error, as Virginia Beach believes, development of approximately 30 mgd of capacity would require either development of added capacity at a future date or use of groundwater. Without institutional reforms, both actions would probably reintroduce water conflicts similar to those that have occurred over the past decade. The excess capacity provided by the pipeline is insurance against future conflicts if projection errors exist.

Approaches to Institutional Reform

Summary of the Problem

Even in states like Virginia where there is no general water scarcity, it is not uncommon for urban areas to look beyond their borders to supplement their water supplies. But the study of the Virginia Beach experience suggests that securing water from other jurisdictions can be an uncertain and difficult undertaking and that institutional considerations are likely to be more important than costs.

Justification for a transfer on efficiency grounds requires that benefits exceed costs. In principle, if the net benefits of a transfer are positive, it should be possible to make all affected parties better off. In practice, however, the benefits tend to be concentrated within the importing area. The water-exporting area is likely to incur some uncompensated, indirect costs such as decreased value of individually held water rights, diminished regional economic activity, and environmental damages. The uncertainties surrounding such indirect costs, and the lack of effective institutional means of providing compensation for them, underlie many of the problems of arranging for transborder water transfers.

The Virginia Beach case study suggests that three conditions, none of them furthered by the Virginia institutional structure, are important for negotiating voluntary water transfers among affected parties. First, property rights to water must be assured—that is, quantified and exclusively defined—and transferable; second, transaction costs for bargaining and reaching an agreement must be low; and third, there must be adequate information about the scope and effects of the proposed transfer. In the absence of these conditions, conflict among groups is to be expected, especially among political jurisdictions that have any interest in a water transfer proposal. Without prospects for low-cost resolution through negotiation, the urban water solution with the greatest net benefits may be abandoned in favor of a solution that promises to generate the least conflict and offers the greatest likelihood of implementation. It was on this basis that the Lake Gaston pipeline proposal was selected. But the same institutional shortcomings that led Virginia Beach to abandon the conjunctive use alternative are proving to be an obstacle for approval of the pipeline proposal as well.

Property Rights. Well-defined, transferable property rights are prerequisite to the easy exchange of property, whether through markets or negotiated settlement. If property rights are not quantified and defined as exclusive property of individual owners, there is risk that payments to secure a right will not result in a secure legal title. The effect of this risk is to reduce the expected value of the right and the incentive to negotiate. Moreover, water rights must be transferable as property so that they may move among owners under mutually agreeable terms of trade.

The Virginia Beach case demonstrates that there is substantial uncertainty about property rights in both surface and groundwater. Surface-water rights, in principle, may be exercised only if the water is used on riparian land. However, case law in Virginia suggests that transfer of water away from riparian land for public water supply is acceptable under the riparian doctrine if other riparian rights holders cannot demonstrate that the transfer has caused them harm. Nonetheless, an investment by a public water supplier to transfer surface water may face a court challenge.

Rights to use groundwater for public water supply are even more uncertain. There is a scarcity of case law to establish a clear interpretation of common law groundwater rights. In addition, certain areas of the states are designated as groundwater management areas. In those areas the Groundwater Act directs the Virginia Water Control Board to issue permits for groundwater withdrawals to protect the future yield of the aquifer and to prevent damages to existing well systems. However, there are numerous ambiguities in the act that confuse rather than clarify the rights to use groundwater for urban water supply.

Currently, agricultural water supply and water for domestic use, including municipal supply, are considered exempt from the permit requirements of the act. Consequently, the state's common law doctrine defines water rights for the exempted uses, while the permitted withdrawals have rights defined by the terms of the act. At this time, neither the courts nor the legislature has clarified the relative priority of permit and common law rights when rights come into conflict. This creates a high degree of uncertainty about any public water supplier's right to use groundwater within a designated management area.

High Costs of Negotiation. As the number of claims on a given volume of water increases, the costs of identifying and negotiating with the owners rise. At some point, the costs of negotiating with multiple owners or claimants will exceed the potential benefits of the transfer, and an alternative that promises lower costs for reaching agreement will be chosen.

Virginia's common law system of rights depends upon private negotiations supplemented by judicial proceedings to resolve conflicts. This system is a low-cost means of resolving occasional conflicts between few individuals over small volumes of water. But in large-volume water transfers, like that associated with urban water supply, these conflict-resolution procedures can be quite costly for all parties involved. Therefore, successful negotiations over a large water transfer may be difficult to achieve in the current institutional setting. Such negotiations can be further confounded because large-scale transfers are subjected to an array of local, state, and federal regulatory procedures that further raise the cost of reaching a decision. For example, the prospects of protracted, costly conflict of uncertain outcome deterred Virginia Beach from pursuing the conjunctive use alternative.

Adequacy of Technical Information. Good-faith negotiation requires adequate information so that the parties can reach a shared, accurate viewpoint about the value of the object of negotiation. In the Virginia Beach case, the best example of the problems posed by inadequate information is the continuing conflict over groundwater development; in part, this is attributable to a lack of generally accepted information about the yield of the aquifers and the effect of groundwater pumping on existing wells. In addition, the opposition to the pipeline proposal has been, in part, a lack of agreement on the effect of a fixed withdrawal on lake levels and instream flows. The inadequacy of information has encouraged "worst case" assumptions in considering the impact of water use on the environment and future economic growth, or evaluating the yield of existing water systems. Further, the water needs of both the importing and exporting areas tend to be exaggerated, along with damage claims associated with any transfer proposal. The result has been a disincentive to negotiation.

Proposals for Reform

Groundwater Rights. Several changes are required in the Groundwater Act to clarify use rights and improve the prospects for negotiating terms for developing groundwater as a public water supply. Extending permit coverage under the Groundwater Act to all withdrawals made in the designated management areas would help clarify groundwater use rights. Common law rights would be superseded. In addition, the Water Control Board should be granted extended powers to regulate groundwater use. As long as the future yield of the groundwater source is not threatened, the Board should have the authority to permit withdrawals even where artesian pressures are adversely affected. Once the Water Control Board determines that further increases in withdrawal rates are inadvisable, permits should be transferable to ensure allocation to the highest valued uses. To protect the rights of all users, the act should specify that parties adversely affected by new withdrawals are entitled to compensation. Procedures for documenting damage claims before the State Water Board, with provision for court appeals on procedural grounds, will also be necessary.

If the act had included these provisions, negotiations would have allowed Virginia Beach to pursue the conjunctive use option and improve its chances of acceptance. Indeed, the Virginia Beach experience has made the state sensitive to the problems with the existing act. Proposals for reform are under study by a legislative committee, and the governor has stated that a plan to revise the act will be submitted to the 1986 session of the legislature. Whether the reform proposal will reflect the suggestions made above is not clear at this time.

Water Markets. A frequent recommendation for institutional reform is for states to make the institutional modifications necessary for the development and operation of water rights markets. Underlying this recommendation is the recognition that water rights markets offer the potential for allocating water resources through interparty negotiation. A second, but no less important, contribution of water rights markets is that continuous trade in water rights generates prices that indicate the value of water in alternative uses.

The quantification, assignment, and recording of rights are prerequisites to the development of water rights markets. Then, for the market to function, an administrative process is needed to monitor and record transfers and to protect certain public trust values such as maintenance of minimum instream flows.

Creation of a water rights market would address many of the problems that have been identified in Virginia's institutional structure. However, institutional change should be subjected to a benefit-cost comparison before it is proposed. The establishment and operation of a water rights market would require a significant expenditure of state funds. In addition, the sharp transition from a common law system of water allocation to a water rights market can be achieved only after protracted political debate. In fact, the legislature has consistently rejected proposals to quantify the water rights into withdrawal permits, an essential step in creating a water rights market (Virginia State Water Study Commission, 1981).

Because the costs would be high, the expected benefits must also be high to justify instituting water rights markets. In Virginia, however, the benefits will be modest. Except for groundwater in the state's designated management areas, water is not an economically scarce resource (Shabman and coauthors, 1981). Indeed, the limited case law adjudicating individual water rights claims is testament to the general abundance of water in the state. Therefore, if a market were established, few trades would occur. The only foreseeable water conflicts that might be avoided through water rights markets are those between political jurisdictions over the expansion of public water systems. In the case of Virginia Beach, cost savings on the order of $60 million might have been realized if market exchanges were possible. But for the foreseeable future, only a few cities in the state are likely to seek expansion of their water systems. Even if similar cost savings were realized in each case, the aggregate savings do not appear significant enough to warrant the monetary and political costs that would accompany the transition to a water rights market. Also, because of this "thin" market, the number of trades would be insufficient to accurately measure the exchange value of water.

Water Transfer Permit Authority. Even though markets do not appear justified, Virginia needs an institutional capability to resolve the disputes that do arise over large-volume transfers of water for public systems. Negotiated market-like settlements remain a desirable objective. Toward this end, the state should institute a water transfer permit authority (hereafter referred to as "the authority") to approve all proposed transfers of untreated water across local political boundaries for public use. The authority's intent would be to facilitate negotiated solutions to water supply conflicts by (1) clarifying the property rights of public water systems, and (2) lowering costs of negotiation between parties affected by a transfer.

There is precedent for creating such an authority in Virginia. Indeed, the creation of special bodies to resolve local government conflicts has been a favored solution of the Virginia legislature and has been quite successful in practice (Richman, 1983). The Commission on Local Government mediates interjurisdictional disputes such as proposals of cities to annex land in adjoining counties. The recently created Hazardous Waste Facility Siting Board is a new effort to facilitate local government negotiations over siting of waste disposal facilities, but it also has the power to make siting decisions that supersede any local ordinance. In the water resources area, a three-judge special court exists to resolve local government disputes over water project development. However, its authority does not extend to water rights adjudication, so its ability to resolve water conflict is limited.

Issuing permits for interjurisdictional transfers of water could be a new authority for the State Water Control Board, an expanded function of the Commission on Local Government, a role for the existing three-judge special court with expanded authority, or the responsibility of a newly formed organization. Because the authority would operate only when infrequent disputes arise, effective use of state resources suggests assigning an existing agency as the permit authority.

The authority would act as an adjudicatory body for resolving conflicting claims associated with a transfer. Current water rights holders, environmental interests, and local governments would all have standing to present claims before the issuance of a permit. The greatest potential for the authority is as a forum for interparty negotiation. Therefore, the authority should seek to mediate disputes and then ratify negotiated solutions by issuing a transfer permit. If a negotiated solution cannot be achieved, the authority would be empowered to impose a legally binding solution on all parties in accordance with prescribed guidelines.

The ability to impose a binding solution is especially important as an inducement to good-faith negotiation when multiple parties are involved (Raiffa, 1982). A useful analogy for this process is provided by legislative provisions for labor-management relations in the federal civil service. That procedure calls for contending parties to call for either binding arbitration or the services of a special impasse panel to mediate the dispute. The panel is empowered to take action to resolve any impasse (5 U.S.C. 7119 (1982)).

The permit would be issued if the transfer promises to provide net benefits to the state and appropriate compensation will be paid by the party making the transfer. Based on the negotiation process, the authority would specify the form (monetary or nonmonetary) and recipient of the damage compensation. For example, funds for environmental damage mitigation could be provided to the state's fish and game management agency. In addition to compensation for damages, the area of the water's origin should receive a share of the net gain from the transfer. This payment could be in the form of a direct payment or, alternatively, of services provided by the transferrer, such as providing treated water.

To prevent continuing conflicts, a water transfer permit should have certain features. First, it should grant a legal right to transfer water from a prescribed location at an established rate whenever streamflows are above a minimum level. The right should supersede all existing common law water rights and regulatory measures of local government. The exercise of the right should be immune to any challenges brought in the state courts or in administrative agencies of the state. (The issuance of the permit could be appealed to the state courts only on grounds of violation of administrative procedure. For a discussion of the constitutionality of this permit authority, see Cox and Shabman, 1984.)

By consolidating decision-making procedures and carefully specifying claimants, the authority would resolve disputes over a water transfer in a single process, substantially reducing the cost of negotiations. In particular, the authority's review process would facilitate compliance with federal regulatory procedures and require mitigating actions for any environmental damages. Local governments would also have standing in this process to oppose transfers with demonstrably negative effects on economic growth. No forum to consider those arguments now exists. Finally, the permit process offers an opportunity to acknowledge the concerns of other states. In fact, opposition to the Lake Gaston pipeline by the state of North Carolina continues at this time. If the permit authority were functioning, testimony from that state's officials and subsequent debate within the review process might have resulted in setting the terms of the permit to address North Carolina's concerns. Of course, if the scope of the interstate conflict includes fundamental concerns over the allocation of interstate waters, then only the federal courts or the creation of a river basin compact can settle the conflict.

The permit process might also sharpen the quality of information used to justify water transfers. By subjecting the factual claims of state and local governments and private parties to open debate and independent review, the authority's review process will encourage the presentation of valid claims.

The most important contribution of the authority would be to increase the certainty of water rights for both public water systems and other interested parties. Permits issued in exchange for compensation payments would assure public water suppliers of exclusive rights to a specific quantity of water. The authority would also ensure expeditious review of the damage claims of water rights holders in the area of origin. Payments for damages would be established before the transfer occurs and would be awarded in accordance with prescribed rules for evaluating damage claims. This differs from the present situation, where damage awards can be made only after a transfer has occurred and has been reviewed in a long court process of uncertain outcome. Moreover, with the proposed authority, the compensation payments made to the area of origin would include a share of the net gains from the transfer. Although payments in excess of damages would be made to local governments, all the citizens of the area would benefit from increased local revenues to supplement tax collections.

Improved Information Base. When conflict is traceable to disagreements of analysis, additional examination is often sought by a third party. In southeastern Virginia the U.S. Army Corps of Engineers was relied upon to provide much of the technical analysis. However, this planning assistance followed federal guidelines and was often delayed by changes in federal policy. As a result, the Corps' work was viewed with some skepticism by local governments and was often not available at critical times in the planning process. Throughout this period, the Virginia State Water Control Board remained almost silent. The state could have acted to address the necessary technical issues without being encumbered by federal planning limitations, but it failed to do so.

This absence of state action prompted the legislature in 1981 to pass a bill calling on the Water Control Board to expand its water resources data collection and analysis capability. The state was seeking to provide definitive answers to technical questions about the state's water resources as they arose, as well as providing technical assistance in the conduct of studies. The intent is to have the state act to limit disputes over technical matters. The state effort is still in its formative stage, so the prospect for success remains uncertain.

Conclusion

Within the current institutional setting, Virginia Beach has chosen the most appropriate solution for its water supply problem. But this solution appears to be more costly than necessary, and the selection has been accompanied by more than a decade of interjurisdictional conflict that is expected to continue. In Virginia, the source-of-water conflict is a selective one—transfer of water for municipal water supply in an environment of general abundance. The southeastern Virginia case illustrates that water transfer conflicts arise and persist because of institutional inadequacies that could be remedied by reforms tailored to the specific nature of the state's water problem. There is no water supply crisis in Virginia. Modest reforms of water resources institutions promise to reduce the currently high cost of conflict resolution and discourage the excessive investments now made to avoid such conflict.

References

Anderson, W. 1978. *"An Economic Approach to Water Supply Planning in Southeastern Virginia"* (Ph.D. dissertation, Virginia Polytechnic Institute and State University, Blacksburg, Va.).

Bauman, D., J. Boland, and J. Sims. 1980. *The Evaluation of Water Conservation for Municipal and Industrial Water Supply-Procedures Manual* Contract Report 80-1 (Fort Belvoir, Va., U.S. Army Engineer Institute for Water Resources).

City of Virginia Beach. 1981. *Water Position Paper* (Virginia Beach, Va.).

———. 1982a. *Lake Gaston Water Resource Development Program—Informational Document* (Virginia Beach, Va.).

———. 1982b. *Water Resource Development Program for Tidewater Virginia* prepared for North Carolina-Virginia Water Resources Management Committee, December 14, 1982.

———. 1983. *Lake Gaston Water Supply Project Environmental Report* (Virginia Beach, Va.).

Cox, W., and L. Shabman. 1983. *Institutional Issues Affecting Water Supply Development: Illustrations from Southeastern Virginia*, Bulletin 138 (Blacksburg, Va., Virginia Water Resources Research Center).

———. 1984. "Virginia's Water Law: Resolving the Interjurisdictional Transfer Issue," *Virginia Journal of Natural Resources Law* (Winter).

Davis, R.V. 1981. "Water Crisis: Working Together Will Give Answers," *Virginia Water*, vol. 8, no. 2 (Richmond, Va., Virginia State Water Control Board).

Engineering News Record. 1971. (New York: McGraw Hill). December 16.

———. 1982. (New York: McGraw Hill). December 9.

Geherity and Miller, Inc. 1979. *Availability of Groundwater in the Southeastern Virginia Groundwater Management Area*. Report prepared for the Virginia State Water Study Commission (Richmond, Va.).

Henningson, Durham, and Richardson. 1975. *Water Sources for Southeastern Virginia*: A Development Plan Prepared for the Southeastern Water Authority of Virginia (Norfolk, Va.).

Hrezo, M. 1981. *Norfolk vs. Suffolk: Proposed Agreement Leaves Important Issues Unsettled*, Special Report No. 14, (Blacksburg, Va., Virginia Water Resources Research Center).

Linsley, R., and J. Franzini. 1979. *Water Resources Engineering* (New York, McGraw Hill).
Maguire, C.E. 1982. *Phase I Study—Roanoke River Basin Water Resource Development Plan* (Norfolk, Va., C.E. Maguire, Inc.).
Prinderville, S. 1983. Memorandum to the Virginia members of the North Carolina-Virginia Water Resource Management Committee, May 2.
Raiffa, H. 1982. *The Art and Science of Negotiation* (Boston: Harvard University Press).
Richman, R. 1983. "Structuring Interjurisdictional Negotiation: Virginia's Use of Mediation in Annexation Disputes," *Resolve* (Summer).
Shabman, L., W. Cox, and D. Ledvina. 1981. *An Assessment of Water Use and Availability in Water Resource Regions of Virginia Through the Year 2000*, Research Report 42, Department of Agricultural Economics, Virginia Polytechnic Institute and State University, Blacksburg, Va.
Sheer, D. 1983. "Water Supply," *Civil Engineering Magazine* (June).
———. n.d. *Assured Water Supply for the Washington Metropolitan Area* (Rockville, Md., Interstate Commission on the Potomac River Basin).
———, and N. Ehrlich. 1982. "Groundwater Recharge for Tidewater Virginia Water Supply" (Rockville, Md., Interstate Commission on the Potomac River Basin).
U.S. Army Corps of Engineers, Norfolk District. 1983. "Yield Analysis—Hampton Roads Area" (Norfolk, Va.).
U.S. Senate Committee on Public Works. 1974. Resolution adopted June 11, 1974.
Virginia State Water Control Board. 1981. *Plan for the Development of the Water Resources of Southeastern Virginia* (Richmond, Va.).
———. 1981. *Report of the State Water Study Commission to the Governor and General Assembly of Virginia*. Senate Document no. 15 (Richmond, Va.).
———. 1983. Historical Review in Current Status of Planning in Southeastern Virginia. Information Bulletin No. 554 (Richmond, Va., Virginia State Water Control Board).
Whitman, Requardt, and Associates. 1971. *Norfolk Water Supply* (Baltimore, Md.).

6
Innovations in Water Management: Lessons from the Colorado-Big Thompson Project and Northern Colorado Water Conservancy District

*Charles W. Howe, Dennis R. Schurmeier, and William D. Shaw, Jr.**

The Colorado-Big Thompson Project (C-BT) that transfers water from the western slopes of the Rocky Mountains to northeastern Colorado both required and inspired institutional innovation. An arrangement had to be devised that could not only operate a complex hydraulic system but could also guarantee the repayment of project costs as required by Bureau of Reclamation laws, negotiate solutions to conflicts with the basin of origin, and allocate water among users with varied water needs. These challenges led to the establishment in 1937 of the Northern Colorado Water Conservancy District (NCWCD) (Dille, 1958; Hartman and Seastone, 1970).

Although the creation of a reliable water supply would have benefited the region under any circumstances, much credit for the success of the C-BT Project must be given to the unusual system of water markets that evolved within the district. In part by direction and in part by chance, a set of legal and administrative conditions was created that made these markets possible. These markets, while having some shortcomings in economic efficiency, are more efficient than those methods

---

*Respectively, professor of economics and Ph.D. candidates, Department of Economics, University of Colorado, Boulder. The authors acknowledge the enthusiastic support and cooperation of the staff and directors of the Northern Colorado Water Conservancy District. City and rural water officials and personnel of the Bureau of Reclamation provided data and advice. Staff support was provided by Marjorie Urban at the University of Colorado and Onengan Soetatwo at Gadjah Mada University, Indonesia.

currently used by federal and state water agencies. These arrangements for allocating water among competing and ever-changing uses might be copied for many places in the West and elsewhere.

The purpose of this chapter is to describe the physical and historical setting in which these institutions evolved, to discuss the unique short-term and long-term market arrangements for the allocation of water that have evolved under the district's direction, to demonstrate the comparative economic efficiency of these arrangements, and to emphasize the potential for wider application.

Evolution of the Project

Physical Setting

If one were to start at Boulder, Colorado, and draw one line straight north to the Wyoming border and another straight east to the Nebraska state line, the resulting rectangle in northeastern Colorado would include all of the lands directly served by the C-BT Project. (The boundaries of the NCWCD are shown in figure 6-1.) The NCWCD region is semi-arid, with average annual precipitation of 13.7 inches. The long-term average annual runoff within the boundaries of the district is 1.1 million acre-feet. Since 1957, the project has provided an average of 230,000 acre-feet, or about 17 percent of the total water supply of the region. This supply is used primarily for supplemental irrigation. However, several towns and a growing number of nonagricultural industries use C-BT as a raw water supply. C-BT water, like the supplies from local streams and aquifers, is of high quality. Therefore, differences in water quality are not an issue in water allocation.

On the western side of the Rocky Mountains, a series of reservoirs captures part of the flow of the Colorado River and its tributaries. Water is pumped up to Grand Lake to an elevation high enough to allow the water to flow by gravity through the Adams Tunnel to the eastern side of the mountains where, at several locations, hydroelectric power, totaling about 250 million kWh, is produced annually. At lower elevations, the water is channeled into the main northern and southern feeder canals for distribution. Natural streams—notably the Big Thompson River—are used as part of the distribution system. Green Mountain Reservoir on the Blue River on the western side of the mountains is not part of the physical system but was constructed to hold replacement water for water diverted from the Colorado River for C-BT. The federal government retains and operates all the collection works and electric power features, whereas the district must provide for the "perpetual care,

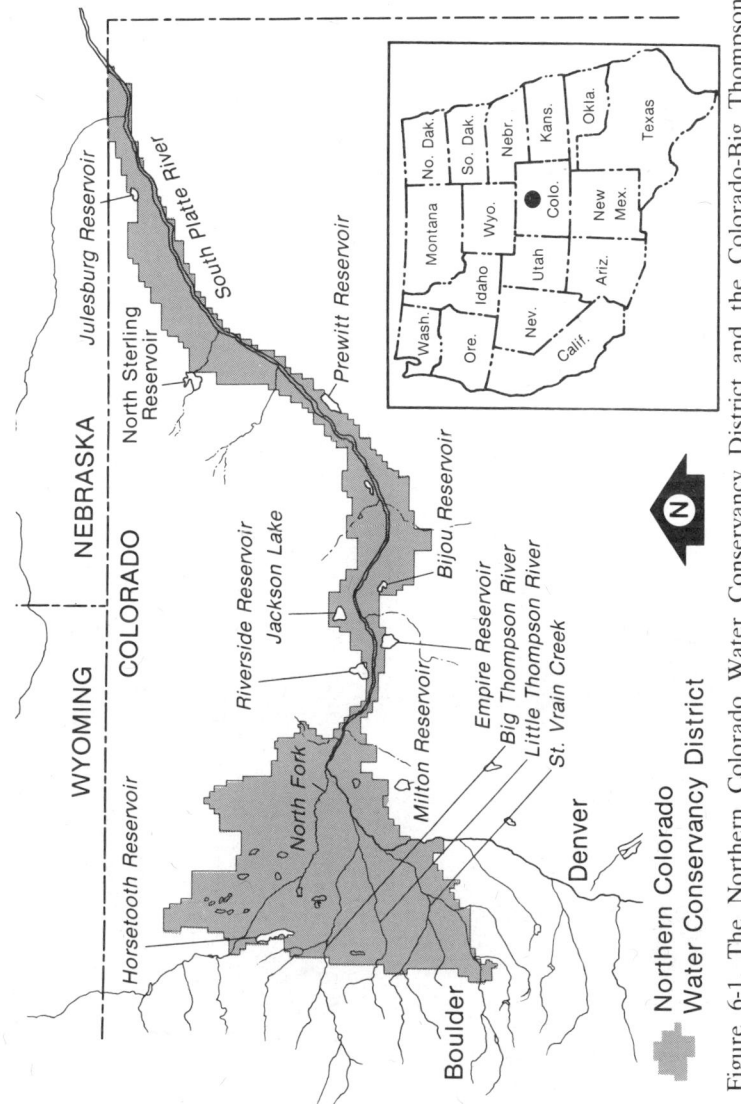

Figure 6-1. The Northern Colorado Water Conservancy District and the Colorado-Big Thompson Project. *Source:* Adapted with permission of the Northern Colorado Water Conservancy District, Loveland, Colo.

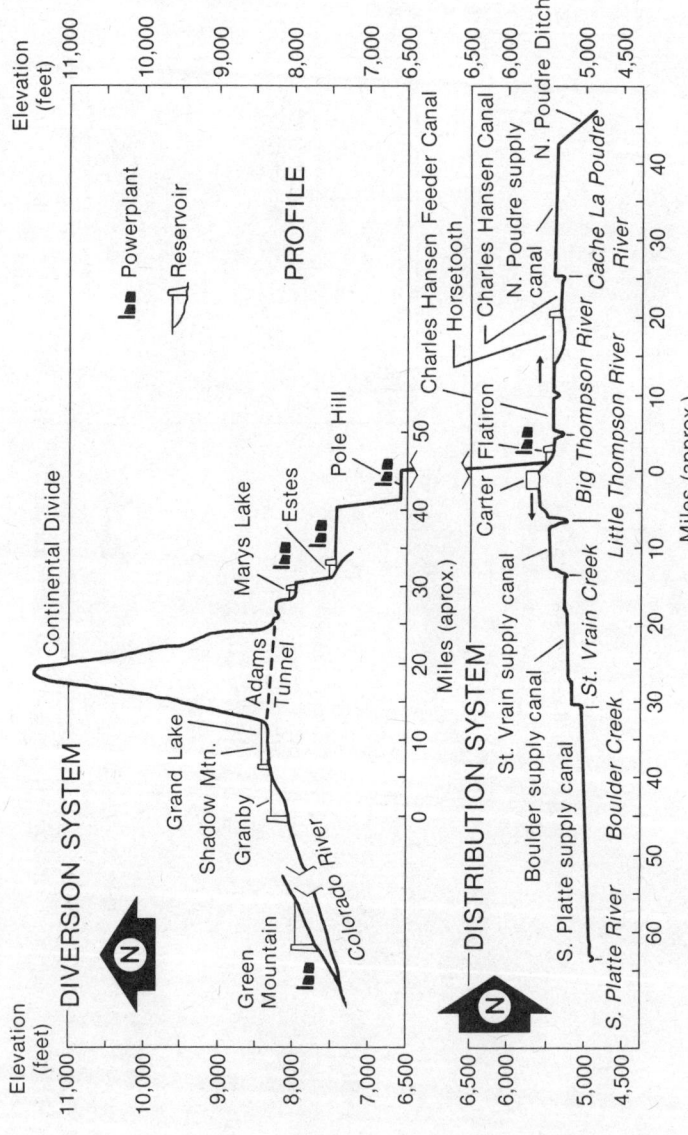

Figure 6-2. Transmountain diversion features of the Colorado-Big Thompson Project. *Source:* J.M. Dille, *A Brief History of the Northern Colorado Water Conservancy District and the Colorado-Big Thompson Project* (Loveland, Colo., NCWCD, 1958) p. 72.

operation, and maintenance" of all the water distribution works. The main physical features of the C-BT are shown in figure 6-2.

Before the allotment procedures developed by the district are described, it would be well to turn to a brief history of C-BT and NCWCD, emphasizing conflict resolution with water users in the Upper Colorado basin and the financial arrangements between the district and the Bureau of Reclamation.

Historical Setting

Construction of the C-BT Project began in 1938, but the idea of transmountain diversion goes back to the 1870s, when the early appropriators of water began to suffer from expanding upstream diversions. The concept of diverting water from the western side of the front rampart of the Rocky Mountains was advanced by nature's heavier rainfall and snowpack on the western side of the Continental Divide in the headwaters of the Laramie, North Platte, and Colorado rivers. Little of that water was being used in Colorado, but few opportunities existed for capturing water on the western side of the divide and diverting it by gravity to the eastern slope. Such projects required a high western catchment basin close to a low transdivide pass at the head of an eastern slope water-using area.

The surge of early diversions occurred from 1890 to 1910 (Cole, 1948). Tipton (1933) mentioned twelve operational transmountain diversions supplying more than 47,000 acre-feet annually from the Laramie and Colorado rivers to the South Platte basin.

The first Colorado state engineer made a preliminary survey of opportunities for finding low passes that could command a sizable runoff from west of the divide. In 1899, the Colorado State Legislature appropriated $25,000 to conduct an investigation of a project involving a canal from Grand Lake to Monarch Lake on the western side, and a tunnel from Monarch Lake to the St. Vrain River on the eastern slope (U.S. Department of the Interior, 1968). In 1904, the Reclamation Service (now the Bureau of Reclamation) began to study a project that involved raising the level of Grand Lake by 20 feet and diverting water 12 miles by tunnel from the lake to either the Big Thompson River or the St. Vrain River (U.S. Department of the Interior, 1968). It is questionable whether the tunneling technology of the day could have accomplished the feat. In any event, no action was taken to develop either project. In 1909, the possibility of bringing Grand Lake water into the Big Thompson by means of a tunnel was again investigated, at which time it was estimated that 500,000 acre-feet could be captured annually in the Grand Lake area.

In 1915, Colorado ceded the area of Rocky Mountain Park to the National Park Service. The January 1915 federal act that created the park contained a reservation permitting the Reclamation Service to "enter upon and utilize for flowage or other purposes any area within said park which may be necessary for the development and maintenance of a government reclamation project" (Dille, 1958). This provision was to become a point of great controversy, but it helped resolve later conflicts over C-BT between the Reclamation Service on one side and the National Park Service and environmental organizations on the other.

The Colorado River Compact of 1922 (Howe and Murphy, 1981) contained a provision that permitted state-allotted water under the compact to be used anyplace in the state, including areas outside the Colorado basin itself. This provision, in effect, foreclosed future interstate (although not intrastate) controversy over out-of-basin transfers.

The period from 1927 to 1937 was one of diminishing streamflow. Drought prevailed in eastern and southern Colorado from May 1931 through 1935, and similar conditions were experienced on the western side of the divide, as shown by the data on Colorado River flows in table 6-1.

In 1934, the Bureau of Reclamation estimated that 75 percent of the 615,000 acres potentially served by the C-BT had inadequate irrigation water supply, presumably meaning insufficient supply to maximize physical yield (U.S. Congress, 1937b). In August 1933, the Colorado administrator of the newly established federal Public Works Administration (PWA) met with the Weld County Commissioners to identify a project that would help the region economically. The possibility of a Grand Lake diversion project was revived. Private parties from the Greeley area supported the proposal, with the publisher of the Greeley Tribune taking the lead in publicizing the proposal. The State Engineers' Office commissioned the deputy engineer, R. J. Tipton, to prepare a

TABLE 6-1. Estimated Mean Annual Virgin Flows at Lee Ferry, Arizona

Period	Mean annual flow (millions of acre-feet)	Characteristics
1896–1929	16.8	34-year wet period
1930–1968	13.0	38-year dry period
1914–1923	18.8	10-year wettest period
1931–1940	11.8	10-year driest period
1934	5.6	smallest 1-year flow

Source: J. A. Dracup, "Impact on the Colorado River Basin and Southwest Water Supply," in *Climate, Climatic Change, and Water Supply* (Washington, D.C., National Academy of Sciences, 1977).

preliminary feasibility report. Known as the Tipton Report (Tipton, 1933), it was a valuable instrument in the early attempts to sell C-BT.

In early 1934, it was decided that acceptance and funding of C-BT as a Bureau of Reclamation project was a superior approach to that of PWA funding, and in 1935, PWA provided $150,000 for a detailed project survey by the Bureau of Reclamation.

By 1934, it became clear that a permanent organization to represent the water users of the region was needed. A committee of eminent citizens representing regional water interests moved to establish the Northern Colorado Water Users Association, which was incorporated in December 1934. This association carried the promotional and organizational burdens of C-BT for the next three years. Funding was voluntary, with ditch companies, county governments, and the railroads as heavy subscribers. The association requested a contribution of one cent per irrigated acre from farmers. The association's budget for the three-year period totaled only $25,412.

The major association activity was counteracting strong opposition to the project by western slope interests who perceived danger to future development from the transmountain diversion of water. Edward T. Taylor, chairman of the U.S. House of Representatives Appropriations Committee, became the champion of western slope interests.

The Bureau of Reclamation reported favorably on the C-BT Project in 1936. A permanent organization with broad administrative and fiscal powers was then required to contract with the federal government regarding project management and cost repayment. At the time, Colorado law had no provision for such an entity. It was necessary, therefore, to allow for the establishment of an entity able to comply with all aspects of reclamation law and, at the same time, be responsive to local needs. This would be difficult, for at least two reasons. First, there were vast differences in local water supplies among the lands of the future district. Reclamation tradition called for a uniform and permanent assignment of project water to the lands. Some flexibility was needed for allocation of water within the project's service area. Second, it was felt that benefits would accrue to the entire area, not just to the irrigated lands, and that equity required the general public to bear part of the project costs. This would require a taxing power.

Accordingly, in 1937 the Colorado Conservancy District Act was introduced (Colorado House Bill No. 714) and signed into law in May of that year. This innovative piece of legislation has since been copied by most western states. Among its major features, the board of directors was given power to acquire and hold property, including water rights; power of eminent domain and to construct works on state lands; power to contract with the United States; power to fix assessments against the recipients of water, according to amount of water received, all rates being "equitable but not necessarily equal"; power to maintain offices and employ needed labor; and power to adopt plans and specifications. The district was also empowered to levy taxes on all real and personal property.

Organization of a district under this act was the next task, the first step being to define the boundaries of a logical area—presumably the one to be benefited by C-BT and in which taxpayers would be asked to bear part of the project costs. It was then necessary to obtain the agreement of at least 1,500 owners of irrigated land and 1,000 owners of nonirrigated or city property on petitions requesting the District Court of Weld County to organize the district. No protests were filed; the court issued the decree establishing NCWCD on September 28, 1937.

Resolution of Conflicts

Sharp conflicts arose over C-BT, with the district pitted against western slope interests on one hand, and against the U.S. Park Service and allied environmental groups on the other. In 1933, contacts with western slope interests were initiated to avoid misunderstanding and confrontation. The key western slope parties had organized the Colorado River Protective Association to defend against incursions into Colorado River water supplies.

In April 1934, a delegation traveled to Grand Junction to negotiate a plan that would benefit both east and west slopes and save some of the surplus Colorado River waters that "otherwise would forever pass out of the State" (Dille, 1958). The western slope position was a "foot-for-foot" compensatory scheme in which C-BT would be required to construct an acre-foot of new western slope storage capable of being filled by spring streamflows for every acre-foot of average annual diversion to the eastern slope. This would have meant 310,000 acre-feet, a large and expensive amount of storage and seemingly quite excessive, since the Colorado flows were far from maximum use. A stalemate over this compensatory requirement continued until June of 1937 when representatives of the two associations in the company of the Colorado congressional delegation hammered out an agreement. The agreement required construction of Green Mountain Reservoir with a capacity of 152,000 acre-feet, and contained relatively detailed operating procedures; 100,000 acre-feet of storage was to be designated for electric power production, and 52,000 acre-feet was designated for supplemental water supply into the Colorado River. The figure agreed upon for total capacity probably evolved from the characteristics of the reservoir site.

Although the conflicts with the western slope were thus resolved, there were other disputes during the same period. Northern Colorado interests had urged the Bureau to undertake a full engineering survey of conveyance routes for C-BT in 1934. The National Park Service informed the Bureau that no surveys would be allowed in Rocky Mountain National Park. This position was strongly supported by the National Park Association—a coalition of wilderness groups. Interior Secretary Harold Ickes strongly supported this position, even though it was he, as acting public works administrator, who had granted the money for the survey in January 1935. However, because of the project's standing as a Bureau of Reclamation project, the legitimacy of transmountain diversions under the Colorado River Compact, and Bureau access to the park under the park's founding legislation, Secretary Ickes had to agree to the survey.

At the time of the hearings on the bill authorizing the C-BT Project, the opposition came out in full force. It was based on a desire to protect national park lands and to avoid setting precedents for water project incursions into national parks (although the Hetch Hetchy Project in California, started in 1927, had inundated part of Yosemite National Park).

The president of the National Parks Association spoke against setting precedent for "economic exploitation" of the National Parks. The associate director of the park service argued for consideration of other diversion routes and the tradeoffs between costs and consequences involved. The forester of the American Forestry Association read a resolution of that organization in opposition to "every bill in Congress . . . which fails fully to protect the national parks from irrigation reservoirs, power projects, or other industrial uses." The national vice president of the Izaak Walton League read a statement specifically opposing C-BT on grounds of encroachment on the park. The American Planning and Civic Association reiterated its support for the national parks and its concern for possible injury to Rocky Mountain National Park. A statement by Horace M. Albright, former director of the National Park Service, was placed in the record. It said, in part:

> . . . I would like to put myself on record as being vigorously opposed to the passage of this bill. . . . It will set a precedent for future invasion and destruction of the national park system.
>
> . . . The very first construction activities will do irreparable damage to Rocky Mountain National Park. . . . Debris from the long tunnel on both sides of the park will be eyesores for a century.
>
> . . . The next park to be attacked will be Yellowstone. . . .

The hearings closed with statements about the careful design work that had been done to avoid environmental damage and the completeness of eastern and western slope agreement on the project.

Review of opponents' arguments as presented in the hearings indicates a lack of knowledge of the project features, their locations, and the impacts on the Rocky Mountain National Park. Most opposition was based on the fear of setting a precedent for allowing incursions into national parks, even though the intrusions of C-BT had been legalized and limited to Rocky Mountain National Park.

In 1937, after Congress had finally authorized the project and the first appropriation for construction was made, the National Parks Association and others made another attempt to stop the project by getting Secretary Ickes to hold another hearing to listen to protests against the authorizing procedures that had been followed. Ickes again stated his own preferences for the Park Service point of view, but agreed that he had to comply with Congress's decision. On January 4, 1938, he announced that President Franklin Roosevelt had formally approved his findings. Controversy was over and the project commenced.

The Repayment Contract of 1938

The Bureau of Reclamation's 1937 preliminary report indicated a construction cost of $46 million. It concluded that the sale of power and the sale of water at $1.00 per acre-foot would make the project feasible under existing reclamation law relating to repayment. No economic feasibility study was conducted, but estimates were presented of the water "shortage" in the intended project area—575,000 acre-feet. This figure was based on irrigating 615,000 acres and, presumably, maximum yield water applications. Average annual gross crop losses were estimated to be $4,700,000, and the extreme-year gross losses of 1934 were estimated to be $12,400,000. The Bureau estimated that C-BT deliveries of 310,000 acre-feet, plus multiple return flows along the South Platte system, would provide a usable supplemental water supply of 554,500 acre-feet. Only Secretary Ickes voiced some doubt about the "feasibility" of the project, but he seemed to refer to repayment feasibility rather than economic feasibility as represented in benefit-cost analyses.

The district had to enter into a contract with the United States specifying the obligations of both parties. The job of shaping this contract both to satisfy reclamation law and to meet the approval of the district taxpayers and water users required ingenuity. The project was to furnish supplemental water to currently irrigated lands and municipal uses. But there were questions about how to allocate the water and how to limit the financial obligations of the district to win voter approval. After months of negotiation, the resulting draft contract differed from previous repayment contracts with the Bureau. The main articles were concerned with the following points:

- allocation of costs between power and water supply
- repayment by the district of those construction costs allocated to water supply
- average quantity of water to be delivered annually
- power features, which were to be entirely the responsibility of the Bureau
- water supply features on the eastern slope, which were to be the perpetual obligation of the district for operation, maintenance, and replacement
- joint features (for example, all western slope features and the Adams Tunnel), which were to be operated by the Bureau with equal sharing of annual costs
- specification of minimum tax rates on property in the district and minimum annual assessments against water users
- disposition of return flows from diversions of project water
- operating steps to protect western slope interests.

Of these points, the first two were of greatest value to the district. From a long-term economic and operating-efficiency viewpoint, the return-flow provision was by far the most important. Voters overwhelmingly approved the contract in June 1938.

Regarding the first two financial provisions, the contract admitted the arbitrariness of allocating joint costs between project purposes, and then settled on a 50-50 split of construction costs between power and water supply. In keeping with Reclamation law, interest-free repayment was to be over a forty-year period in accordance with an increasing schedule. In addition, an upper limit of $25 million was set on the total repayment obligation. At that time, the estimated construction cost was $44 million, but the actual cost came to $163 million. The value of these provisions to the district was almost surely not appreciated at the beginning.

The importance of the special provisions governing return flows was also clearly not appreciated at the time. Under Colorado water law and appropriation doctrine generally, return flows belong to and are part of the stream, and cannot be claimed by the water rights holder who made the diversion. These return flows are subject to appropriation by downstream parties. Because the Bureau of Reclamation had filed for rights claiming the project waters of the Upper Colorado, and because these waters would be new waters to the streams on the eastern slope, Article 19 established U.S. ownership of all return flows from the project and reserved these return flows for district recapture and use. By preventing the filing of individual state water rights claims against the return flows, this provision greatly increased the flexibility for the allocation of water among users because it relieved buyers and sellers of project water from legal damages to return flow users. The Bureau's reason for claiming ownership of return flows was to allow the district to resell these flows and increase the revenue base. The district felt that such attempts would be impracticable to administer and enforce. (Return flows will be discussed in more detail in the section on NCWCD water markets.)

Finally, the contract incorporated the "Manner of Operation of Project Facilities and Auxiliary Features" (U.S. Congress, 1937b) that detailed the rules to be followed to protect western slope interests. These rules cover the operation of the western slope Green Mountain Reservoir on the Blue River. Interestingly, these rules specified that unneeded Green Mountain water was to be made available for the development of "the shale oil or other industries."

An important provision for C-BT not included in the contract was the exclusion of C-BT from the 160-acre limitation. Since the project provided supplemental irrigation water to already established farms, it was exempted from the Bureau of Reclamation law prohibiting farms in excess of 160 acres from receiving federally supplied water. The elimination of limits on farm size in the use of C-BT water facilitated the free transfer of water among users.

Construction Schedule

In July 1938, an amendment to the Interior Department appropriation bill was passed, providing $900,000 for the start of construction. The following dates and events were the highlights of the construction period:

1938—work starts on Green Mountain dam and power plant
1940—commencement of drilling the 13-mile Alva B. Adams Tunnel, driven simultaneously from both ends
1942—construction stopped by order of the War Production Board, except for completion of the power features of Green Mountain
1943—first power generated at Green Mountain Reservoir
1944—Adams Tunnel holed through
1947—delivery of first water into the Big Thompson River for use along that stream
1950—first generation of power on the eastern slope

1956—construction of the last major water distribution feature, the South Platte Supply Canal
1957—the first year of full project water deliveries.

NCWCD Water Markets

Allocation Rules: Theoretical Discussion

The water markets that evolved from the C-BT Project contain a number of interesting features. Before describing the operating rules and unique features of the NCWCD and presenting empirical evidence from the markets, it would be well to discuss the various theoretical considerations governing different allocation systems.

The major methods for allocating large water supplies are priority allocation rules and proportional rules. Such rules can be imposed by law or contract, or they can be incorporated into a system of property rights in water in which market processes determine the holder of the rights. All systems need an administrative framework to ensure that the rules are consistently observed.

Under a priority rule, the various users are assigned certain quantities of water per time period, and each of these quantities has a priority number. Each period (or each day if the priorities are stated in terms of flow rates) the higher (senior) priority users are allowed to take all of their water before the lower (junior) priority users are allowed to take any. If senior-priority users choose not to take their water, the junior-priority users can claim it.

The advantages of a priority system are that it can assign high priorities to those economic activities that need a reliable supply, whereas activities that are not as sensitive to water shortage can be assigned lower priorities. This assumes, of course, that there is some mechanism for distributing the priorities according to need. If a market for transferable shares is used, then those activities requiring a reliable supply can bid for the senior shares, while others can buy more-junior shares at lower prices. Thus, the user can achieve security and still hold claim to water equal to the average amount needed.

One disadvantage is that strict priority use of water, although it may protect sensitive crops, among other uses, involves some inefficiency during periods of shortage through failure to equate marginal-value products among uses. For example, during a shortage, one farm holding a high priority may receive the same amount of water that it applies when water is plentiful, while another farm gets no water at all. Economic efficiency may require each farm to receive some water. In markets that permit water transfers, this problem may be solved by short-term exchanges of water among the holders of different priorities (for examples, see Howe et al., 1982).

Another problem arises when the priority shares are to be traded in a market. Each share is unique, so the market must arrange trades in dissimilar units. Such markets are more difficult to organize.

A proportional rule divides available water among a group of users according to a fixed set of proportions. One way to organize this system is to have shares, or rights, each entitling the holder to the same fraction of available water (for example, if there are 1,000 shares, each share entitles the holder to 1/1,000 of the available water). Of course, the system does not have to operate with tradable shares. Again, the assignment of shares to users can be implemented by a water authority or through a market process. Under a proportional system, risk hedging is accomplished by holding claims to water in excess of average needs so that the probability of shortage is reduced.

Major advantages of proportional rule are greater administrative simplicity and greater ease in organizing a market for tradable rights when a market mechanism is used. These advantages stem from the uniformity of the rights. Without liability for return-flow effects, all rights are interchangeable, making establishment of an efficient market much easier than is the case with priority rights. A second advantage to the priority rights system occurs when the various water users have similar water demand (benefit) functions. They would be expected to hold similar numbers of rights, so the available water would be divided equally among them. This equal division of water should approximate equality of marginal values of water among users.

A feature of the proportional system is that insurance against shortages during droughts must take the form of accumulating shares, from which the yields far exceed normal use much of the time. This excess water is usually rented back to the agricultural sector on a seasonal basis. These rental supplies entail more risk than owning an allotment because the allotment owner may choose not to rent in some seasons. This additional risk precludes the application of this water to higher-valued perennial crops or industrial uses. Under a priority system, the same certainty could be obtained by buying a smaller number of senior rights. It thus appears that a priority rights system has advantages in systems in which risk avoidance is quite different among users, while proportional rights systems have efficiency advantages when users' demand functions and risk avoidance are similar. Demand functions are likely to be similar in areas devoted primarily to a uniform type of agriculture.

Allocation Rules in the NCWCD

The C-BT Project was intended to provide supplemental water supply to agricultural, municipal, and industrial users. A major issue in establishing NCWCD was the method for distributing the water among eligible users. The area encompassed by NCWCD included areas of quite different natural water supplies in relation to the amount of arable land. Some areas served by ditch companies holding very senior rights had large, reliable supplies, while other areas had inadequate, unreliable supplies. This led to different degrees of enthusiasm for C-BT and to diverse opinions about the appropriate distribution method among water users. One point became clear as public meetings on this issue were held: people did not want a mandatory, uniform assignment of water to the land. They and their ditch companies wanted to be able to decide whether to subscribe to C-BT water. These sentiments finally led, in 1957, to a clear-cut definition of an allotment as a freely transferable contract between the district and the holder, subject to the holder's ability to show beneficial use of the water within the boundaries of the district. Moreover, the available C-BT supply is to be evenly divided among the fixed number of allotments, and thus water allocation within the district is based on a proportional system. The transferable and homogeneous nature of the allotments naturally stimulated the creation of a market in which they could be traded.

The total amount of C-BT water available to the district each year is determined by the so-called quota system. The district board of directors annually determines the quota, the percentage of 310,000 acre-feet that will be requested from the Bureau. (Actual deliveries can fall below the quota if the water is not called for by the individual allotment owners.) Starting in January, the directors study the evolving climate, especially the snow pack on both slopes, and district soil-moisture conditions. At their April meeting, they set the quota at some number that may be as low as 60 percent (186,000 acre-feet) in a wet year, or as high as 100 percent (310,000 acre-feet) in a dry year. Thus, the actual size of a single allotment of C-BT water varies from year to year.

Return Flows in the NCWCD

From an operational viewpoint, the other unique feature of the C-BT-NCWCD system is that the return flows are owned by the district rather than again becoming part of the stream as called for under Colorado law. This claim was possible because all the water was new to the basin.

The significance of the return-flow provision in the 1938 repayment contract stems from the interdependence among water users. When water is diverted and used, part of the water typically returns to the stream, where it can be used by others farther downstream. Whenever a change in location or type of use is made, the pattern of return flows will change, damaging some parties and benefiting others.

Return-flow interdependence has at least two important implications: (1) simple buyer-seller market processes will ignore the impacts on third parties and may result in inefficient transfers; and (2) some legal systems attempting to protect third parties may cause high transactions costs or result in inflexibility in water allocations. These points are relevant to the comparative evaluation of institutions for water allocation.

Under appropriation doctrine, return flows again become part of the stream and are subject to downstream appropriation. Any user's return flow is likely to provide the water that makes it possible for a more-junior appropriator downstream to exercise his right. State laws act to protect such downstream parties from adverse impacts of changes in upstream water use patterns, including transfers among uses, by requiring the advertising of any proposed changes in points of diversion or in end uses. Parties who expect to be damaged by such changes are then permitted to petition the water court (or state engineer in New Mexico) for relief.

This relief can take the form of direct compensation from the new water user, in which case the damaged party agrees not to contest the transfer. This happens frequently when towns buy agricultural water rights, even though large numbers of third parties are involved. Such direct compensation should result in efficient transfers. If no agreement is forthcoming, the court reduces the amount the new user can actually divert to quantities that will eliminate all damage to downstream parties—usually the quantity that has been consumed in the past. The court's decision is based on evidence presented by the two sides. This process can lead to inefficiency in the transfer process by preventing desirable transfers, basing a decision on biased information, and increasing transactions costs.

An important implication of the C-BT return-flow ownership provision is that the NCWCD does not have to be concerned about the return-flow problem, which, as described above, can inhibit transfers. When a transfer of NCWCD water is proposed, no downstream party has legal grounds for objection.

This does not mean, of course, that there are no real impacts on downstream parties who use these return flows. The possibility of inefficient transfers exists because the real third party effects are ignored. Therefore, the question is whether or not the advantages of an easy, low-cost transfer process offset any net adverse effects to third parties.

History and Operation of the Allotment Market

In the NCWCD, when a buyer and a seller wish to effect an allotment transfer, they submit an application to the district; the transfer must be certified by a district field crew to ascertain that beneficial uses of the water can be made. This investigation and the board's approval are intended to act as a check on purely speculative purchases of allotments. Municipal and domestic water company users are generally permitted to buy without such checks, principally because "reasonable beneficial use" is much harder to define than in agriculture. This bias against speculation is a curious part of the culture of agricultural areas. Many farmers have made their fortunes by continuing to irrigate long after farming ceased to be profitable, simply to take advantage of rapidly rising water prices. This type of speculation is probably thought of as being different from simply buying water and not using it because of the pervasiveness of the beneficial-use doctrine under state law. The requirement of showing beneficial use only serves to increase the costs of water transfers.

In recent years, several individuals have been acting as brokers in NCWCD allotments. Two modes of broker operation can be traced through the allotment records: (1) the broker simply finds parties who want C-BT water and others willing to sell; and (2) the broker acts as speculator-seller. In the second case, the broker must own agricultural land to which C-BT water could be applied. When an opportunity to buy allotments arises, they are transferred to the broker, ostensibly for agricultural use. The water may actually be used for awhile, or it may be rented until an acceptable buyer is found.

Urban and nonagricultural industrial growth have been high on the eastern slopes of the Rocky Mountains (the front range) since the mid-1960s. Much of the water needed for this growth has been provided by the transfer of NCWCD allotments from agriculture to these new uses. Although this water has become expensive, its ready transferability and high reliability have made it attractive for these rapidly growing sectors. This is reflected in the changing composition of allotment ownership. Whereas irrigators started in 1957 with more than 85 percent of the allotments, by 1982 their ownership had dropped to about 64 percent.

The share of water delivered for nonagricultural uses also has increased over time. But actual water deliveries to nonfarm users have always lagged well behind ownership of the allotments. In 1980, for example, irrigators used 71 percent of the available water even though they owned only 64 percent of the allotments (Northern Colorado Water Conservancy District, 1980). Cities and multipurpose users are "renting" water back to irrigators in large quantities on a short-term (annual) basis. Because any water an allotment holder owns but does not call for by the end of the season reverts to the district, it makes sense for the owner to rent excess water. (Details of the seasonal transfer or rental market will be examined in the next section.)

The original NCWCD allotments were free. However, the allotment holder had to agree to pay an annual assessment of $1.50 per allotment, whether the water was used or not. Many irrigators in the early years gave away their allotments because they did not want to pay the assessment.

To get price data, it was necessary to ask municipalities (the largest class of net purchasers) in the district to provide a list of their allotment purchases and the corresponding prices. These data do not include all transactions, but these prices were apparently representative for all transactions. The data are shown in table 6-2.

Allotment prices increased from $30 in 1960 to $291 in 1973, a compound annual rate of increase of 19 percent. This trend accelerated sharply, with average prices reaching a peak of $2,161 in 1980—a com-

TABLE 6-2. Some Price Statistics on NCWCD Allotments
(current dollars per share)

Year	Average	Low	High
1960	30	—	—
1961	30	—	—
1962	35	—	—
1963	57	40	60
1964	77	60	100
1965	99	95	100
1966	113	100	125
1967	115	60	300
1968	145	122	150
1969	221	200	225
1970	252	200	300
1971	265	—	—
1972	263	250	270
1973	291	263	310
1974	400	315	420
1975	594	435	875
1976	737	467	1310
1977	835	600	1200
1978	1321	1100	1800
1979	1985	1700	2200
1980	2161	1900	2350
1981	1907	1850	1925
1982	1500	1300	1700
1983	1200	1150	1250

Source: Questionnaires from cities in NCWCD.

pound annual rate of increase of 33 percent since 1973. Between 1980 and 1985, prices have fallen back to about $800.

History and Operation of the Rental Market

Rentals are transfers of water among users for one season only. Allotment owners may decide that they have water that either will not be needed or that will be of only low value, and may try to sell it to another water user. C-BT water can be rented to any water user in the district. Through clever exchanges and replacements, C-BT water can even be effectively transferred to parties not on the NCWCD delivery system.

Rentals occur not only with NCWCD water but also with appropriated water and ditch company water. Rentals of non-C-BT water in and between ditch companies occur frequently. Rental of C-BT water can be accomplished between any two water users in the district, whereas rentals among ditch companies are limited by the physical possibilities of exchange. For the same reasons given for allotments, rentals of C-BT water are unencumbered by any liability for third party effects.

To execute a rental, prospective buyers or sellers let the district office know that water is either available for rental or is needed. As rental inquiries come in, the various parties are put in touch with one another. Contacts among people on the same ditch occur naturally. Infrequently, an auction of rental water will be advertised. In cases other than the auction, prices are simply agreed to by the buyer and seller. Transactions costs are quite low, except in unusual years when rental water is hard to find. Rentals are thus easily effected in response to relatively small discrepancies in water values among users.

Analysis of the data on rentals across the sub-areas of the district indicates: (1) about 30 percent of the C-BT water delivered to the district is involved in rental transactions each year; (2) the agricultural sector is a big net rentee, with net amounts rented showing a slight downward trend in the past decade; and (3) the towns are the big renters of water, with a slight trend toward using more of their allotments each year as growth occurs.

A remaining question about rental patterns concerns the difference between dry and wet years. The outcome is complicated by the fact that with the quota system the NCWCD directors order more water from the Bureau in dry years than in wet years. The experiences of two years have been analyzed: 1977—an exceptionally dry year in the 1976–1978 drought cycle, and 1979—an exceptionally wet year. The characteristics of these years are given in table 6-3.

The volume of rental transactions was much larger in the dry year

TABLE 6-3. Comparison of Wet Year with Dry Year Rentals

	1977	1979
Precipitation (inches)	11.55	19.10
Percent of normal	84	140
Quota (%)	100	60
Actual deliveries (acre-feet)	309,477	144,459
Rentals (acre-feet)	125,130	89,507
Rentals as percent of actual deliveries	40	62
Rentals as percent of historical average delivery	48	34

Source: NCWCD records.

than in the wet year. This was partly a result of the larger water supply: in 1977 310,000 acre-feet (100 percent of quota) was requested and received from the Bureau. Also, in a dry year there is a greater sensitivity to water management and keener competition for available water supplies. Not surprisingly, deliveries started in April in 1977, but not until July in 1979.

Anderson (1961) recorded 1959 rental prices for ditch company water. Although ditch company water is not as transferable as C-BT water, each of the ditch companies studied held NCWCD allotments that also could have been rented. Thus rental prices would be about the same for both types of water. The early-season range was $2.50 to $5.00 per acre-foot, while the late-season range was $2.70 to $8.00. Asmus (1966) gathered similar data for 1966. The early-season range was $3.00 to

TABLE 6-4. Rental Price Summary for C-BT Water
(dollars per acre-foot)

Year	Low	High	Predominant range
1969	—	—	3.50[a]
1970	—	—	4.00[a]
1971	—	—	4.50[a]
1972	—	—	4.50[a]
1973	—	—	4.50[a]
1974	4.50	7.00	—
1975	4.50	7.00	—
1976	5.00	8.00	5.00 to 7.00
1977	5.00	8.00	5.00 to 7.00
1978	5.00	8.50	5.00 to 7.00
1979	5.00	9.00	5.00 to 7.00
1980	5.00	10.00	5.00 to 7.00
1981	5.00	25.00	7.50 to 11.00

Source: Ditch company offices and city water departments.
[a] Record of only one transaction.

$7.50, and the late-season, from $5.00 to $12.00. The average early- and late-season prices thus increased at an annual (compound) rate of 5 percent between 1959 and 1966. During the same period, allotment prices rose at a compound rate of 21 percent.

The present study gathered more data on rentals of C-BT water (summarized in table 6-4). The recorded high prices increased a little from 1974 to 1980, but the predominant range of $5.00 to $7.00 remained the same.

Interestingly, the only true market test of rental prices occurred in 1981 when a small town auctioned off excess C-BT water. It was a sealed-bid auction, and nearly all the bids were substantially above the traditional $5.00 to $7.00 range in table 6-4. The highest bid was $25. Even though these bids express willingness to pay rather than equilibrium price, they suggest that rental prices are below a true equilibrium level.

It is difficult to calculate a unique rate of rental price increase from 1966 to 1981, but it clearly lies between 2.6 percent and 3.1 percent, compounded annually. Given that the 1966–1981 period experienced inflation rates from 2.5 percent to 14 percent, real rental prices did not rise at all over this period.

Several factors tend to depress rental prices far below comparable allotment prices. First, rentals carry a higher risk of nonavailability and greater price uncertainty than allotments. This means that the application of rental water is likely to be restricted to supplemental irrigation or to low-cost, low-value crops. A second factor is the moral suasion exerted by the district and the community at large on renters not to "gouge" the farmer with high rental prices. The district directors have allowed cities to own allotments that are clearly in excess of the current average use, the rationale being that cities must have a reliable supply. The unwritten but often stated *quid pro quo* for this special treatment is an obligation not to profit from rentals. Most towns simply add a small administrative charge (frequently $0.50 per acre-foot) to their variable cost of $5.00 per acre-foot when setting the rental price. There have been instances in which cities tried to raise prices toward a market clearing level and were quickly, if quietly, admonished by the district for doing so. In fact, the directors see expanded city ownership of allotments as a way of providing more cheap rental water to the agricultural sector that, increasingly, cannot afford to apply expensive water to low-value crops.

A third factor that depresses rental prices by increasing the average amount available is the procedure by which the C-BT quota is set. As noted above, the amount of water the district receives from the Bureau varies under the quota system. The quota-setting procedure, focused as it is on agriculture, has the effect of providing less water in wet years and more water in dry years. For cities and industries whose water demands are much less dependent on climate, this means they must hold extra allotments as "protection against wet years" (that is, low quotas). This "hoarding" of infrequently used allotments increases the average supply of rental water.

In summary, there is an active rental market, involving about 30 percent of the C-BT water delivered each year. This large volume of transactions clearly indicates a flexibility in allocation that most water allocation arrangements do not have. Most arrangements (for example, Bureau of Reclamation contracts in the Central Valley Project of California) do not permit or motivate short-term transfers among users.

Costs of Alternative Supplies and the Marginal Value of Agricultural Water

All available water supplies are substitutes for one another to some extent. Sources may differ in the critical characteristics of water quality, reliability, and possible modes of delivery, but, being substitutes to greater or lesser degree, their prices must move together because users can shift from one to another. The substitutes for C-BT water delivered by NCWCD would be ditch company shares, individually owned water rights established under state law, short-term rental of either C-BT water or water from state water rights, or groundwater.

Since NCWCD allotments are uniform, easily transferred, and reliable, it is easier to establish a market in allotments than in the diverse state water rights. For this reason, NCWCD allotments tend to set the price pattern for other transferable water supplies. Figure 6-3 shows the relationship among NCWCD allotment prices, North Poudre Irrigation Ditch Company shares, and the "in-lieu-of rate" that developers are required to pay the cities of the region for new raw water to serve their developments.

On the other hand, NCWCD allotment prices are affected by new water developments. This is most clearly seen in the pattern of allotment prices since mid-1980 and its relationship to the progress of the Windy Gap Project, a supplemental water supply project for towns that will bring an additional 48,000 acre-feet of water from the headwaters of the Colorado River to towns of the eastern slope. This supply is exclusively for towns and power companies and cannot be traded. The participating parties do own their return flows, which can legally be sold, the buyer being allowed to divert a quantity equal to the return-flow rate, without regard to priority. The cost per acre-foot of this water is estimated to be $263, counting no return flows, or $160, counting on 65 percent return flow ($263 ÷ 1.65). As figure 6-3 shows, average

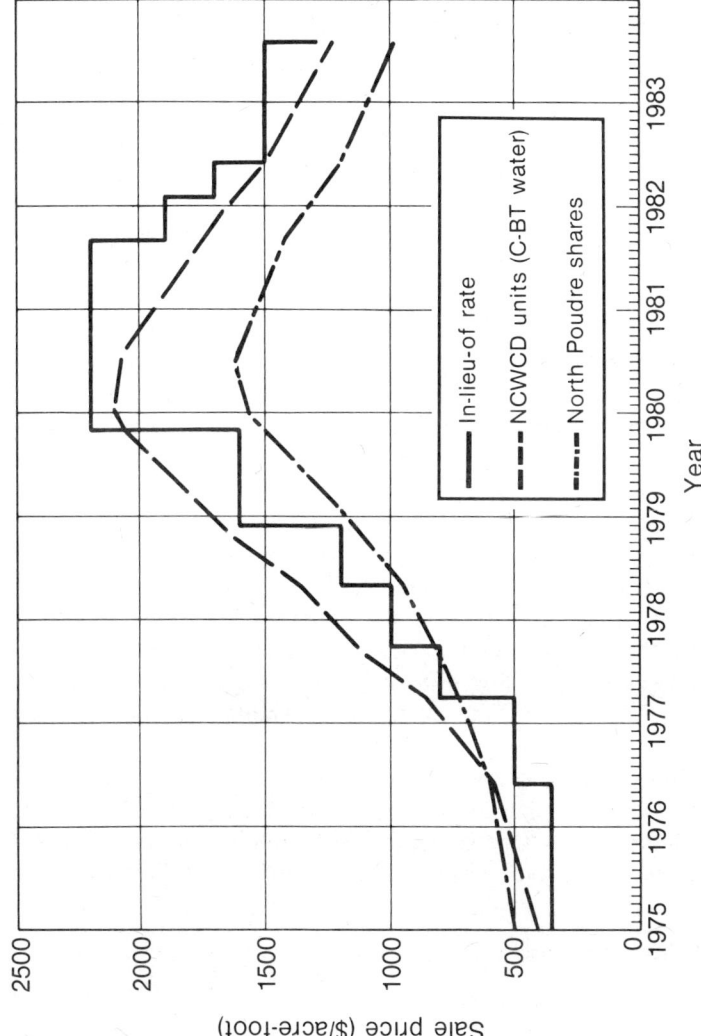

Figure 6-3. Comparison of NCWCD allotment price with ditch company share prices, 1975–1983 *Source:* Fort Collins, Colorado, Water and Sewer Utility

allotment prices peaked at $2,161 in 1980 and fell back to about $1,200 in 1983. The movement in prices since 1980 corresponds in time to the final clearing of several legal and procedural obstacles to construction of Windy Gap and its approaching completion. Amortizing the $1,200 allotment price at 10 percent, and adding the $2.00 annual service charge gives an annual cost of $122 per allotment. Assuming an average of 0.7 acre-feet per allotment, C-BT water costs $174 per acre-foot. Thus allotment prices have fallen almost equal with the newest alternative supply for the buying cities and industries. At the moment, raw water supply and demand seem to be stabilized at these prices. The slightly higher price for NCWCD allotments can be accounted for by their tradability. It seems that some buyers expect allotment prices to increase once again as the region continues to grow.

How can water prices of $160 to $174 per acre-foot be reconciled with the low marginal value of water in irrigated agriculture in northeastern Colorado? Kaleta (1976) has estimated the marginal value of irrigation water in nongroundwater areas to be about $32 per acre-foot. Why would a farmer continue to hold allotments at an opportunity cost of $174 per acre-foot when the marginal value in agriculture is so much lower?

One reason may be the inertia of people who don't understand this opportunity cost or who want to remain in agriculture until retirement. However, many people do retire each year, and prospective buyers of the land and water surely are aware of the opportunity to simply sell the water and allow the land to revert to dryland farming. The more likely reason, then, is an expectation of continued price increase: farmers speculate in water by holding onto it long after its marginal value in irrigation has fallen below its market value. Assuming a marginal value of water of $32 per acre-foot, an allotment price of $1,200, a marginal income tax rate of 40 percent, a capital gains tax of 25 percent, and a discount rate of 10 percent, any farmer anticipating an allotment price increase of more than 4.5 percent can justify holding all his allotments for another year.[1] Given the 10 percent compound rate of increase from

---

[1]The rate of expected price increase that would make it rational for a farmer to continue holding allotments can be computed using the following notation:

t = the farmer's marginal income tax rate
g = the farmer's capital gains tax rate
r_p = the expected annual rate of allotment price increase
r = rate of return on alternative investments, before tax

The after-tax return from keeping and applying the water from an allotment (including anticipated capital gains on the allotments) would have to exceed the anticipated return from selling the allotment at current prices and investing the net proceeds for one

1970 to mid-1983, such an expectation would not be unreasonable. Interestingly, then, the high prices for C-BT water are not inconsistent with the retention of much C-BT water in agriculture.

Efficiency of the NCWCD Water Market Arrangements

Market Versus Nonmarket Arrangements

The use of markets means flexibility over time in the allocation of water. This is extremely important in a growing region with a changing economic structure. Even though individual sales may involve unaccounted third party damages, market flexibility means that the pattern of water allocation will follow the evolving pattern of new, higher-valued uses. Certainly, the NCWCD system is more efficient than the typical Bureau of Reclamation contractual arrangements, which tie water perpetually to the same land and, in many cases, to the same uses. Under many contracts between the Bureau of Reclamation and conservancy districts, water markets are not possible. Inflexibility in patterns of water use like those found in central Arizona and the Central Valley of California either stifle further economic development or require enormously expensive new water projects to supply water for growth. The Central Arizona Project and the California State Water Project are two of the most expensive projects ever undertaken anywhere in the world. The well-known study of the role of water in affecting the growth of the Arizona economy (Kelso and coauthors, 1973) showed decisively that an efficient transfer of relatively small amounts of water out of (low value) agriculture to the newly emerging urban and industrial uses was adequate to maintain rapid state growth without the Central Arizona Project. In a water market system, such transfers take place.

---

year. Using illustrative figures introduced earlier, this can be approximated by:
$$\$32(0.7)(1-t) + \$1{,}200\, r_p\, (1-g) \geq \$\, [1{,}200(1-g)]\, r(1-t)$$

The first expression on the left is the marginal after-tax return in applying the 0.7 acre-feet of water in agriculture. The second expression on the left is the net expected increase in the market value of the allotment, allowing for capital gains taxation. The right side represents the return from investing the net proceeds from selling the allotment.

Assuming $t = 0.4$, $g = 0.25$, $r = 0.10$, the anticipated rate of allotment price increase that should leave the farmer indifferent between keeping and selling his marginal allotment is 4.5 percent.

Importance of a Tradable Margin

Transferability of water raises the concept of a "tradable margin" of water supply. In the case of northeastern Colorado, C-BT water, while representing only 17 percent of the total supply, constitutes an easily transferable margin of supply. The water supply having the lowest transactions costs determines the closeness of the approximation to an efficient allocation. The transactions costs associated with NCWCD allotments and rentals are much lower than those associated with water rights established from natural sources under state law. Thus, C-BT water is the "easily saleable margin" of water and plays a disproportionately important role in the efficiency of water use. As long as such water is widely held throughout the area, economic efficiency will not be seriously impaired by the fact that other sources are nontransferable (infinite transactions costs) or transferable only at high cost (such as state water rights).

Return-Flow Externalities

Return-flow externalities remain a problem and a possible source of inefficiency in any transferable water rights system. Any system that fails to take these external features into account may well generate inefficient transfers of water in which costs exceed benefits. NCWCD, through its ownership of the return flows of C-BT water, has made it possible for any buyer-seller combination to escape liability for return-flow externalities and has, thereby, facilitated market transactions. A necessary question, then, is whether or not the advantages from reducing transactions costs and stimulating market transactions more than offset possible third party net losses.

In the particular geographical setting of NCWCD, there are reasons to expect that net third party effects of allotment sales and rentals are positive (that is, *gains to users downstream from buyers exceed losses to users downstream from sellers*). The initial transfer is likely to be from a less-productive area to a more-productive area. These value differentials are also likely to hold for downstream parties. Moreover, many of the return-flow effects are experienced by parties much farther downstream on the main stem of the South Platte, and these return-flow uses are not dependent on the particular location of the original upstream diversion.

These fortuitous circumstances of C-BT and NCWCD are not likely to be found in other locations, so the procedures for handling return-flow externalities still must be addressed. In general, allocative efficiency requires evaluation of third party effects, and market efficiency in particular requires the establishment of property rights in return flows. Obviously, these property rights must be quantifiable in value and enforceable by law.

Thus, return-flow externalities generally are best treated not by ignoring them (as in NCWCD) but by streamlining the processes: identifying and quantifying the externalities, facilitating compensation to damaged parties, and allowing the buyer to sell return flows. This point was made by Hartman and Seastone (1970) who also described the water transfer system in New Mexico, where the state engineer is called upon to identify and quantify return flows, thereby greatly reducing cost and uncertainty.

Proportional Water Rights

As previously noted, proportional rules for allocating water are particularly suited to conditions where users are similar. In the NCWCD, however, there are two major classes of users whose demand functions and risk avoidance differ considerably: agriculture and urban-industrial users. Under these conditions, some differentiation of rights by priority would be efficient—perhaps having two classes of rights: first priority and second priority, with proportional rights within each class. However, such a system would suffer its own source of inefficiency, because a transfer of rights from agriculture to urban-industrial would create a "risk externality" on all remaining agricultural users by increasing the demands to be satisfied ahead of agricultural demands.

In the case of NCWCD, two factors already noted interfere in an indeterminate way with achieving an efficient allocation of shares between agriculture and the urban-industrial sector. The first is the quota system, which causes variation from year to year in the amount of water per allotment, thus motivating urban-industrial users to acquire more shares to "protect themselves from wet years." The second is the artificially depressed rental prices, which have the effect of increasing the annual cost of holding an allotment. This increase would be zero for a user who never rented out water, although it would be quite high for a city that holds many allotments that are typically rented back to agriculture.

This strongly suggests that NCWCD and other groups contemplating water markets should eliminate the quota on all shares held by urban-industrial users and allow the rental market to operate without constraint. The overall effects would be to increase the efficiency of the proportional rights system, to increase the availability of rental water during drought, and to ensure that parties placing the highest value on rental water actually acquire it.

Conclusions and Recommendations

In the context of NCWCD and its geographical setting, the markets that have evolved appear to be much more efficient than the contractual arrangements that typify Bureau of Reclamation projects, for example, the Central Valley Project. The existing market procedures can clearly be improved, and future adoption of these procedures should avoid the distortions of uniform application of the quota system and the artificial ceilings on rental prices.

For any rapidly changing region, the use of a priority water rights system as the allocative device for new water supplies would seem to have an advantage where degrees of risk avoidance vary greatly among users. The "hoarding" of proportional shares to hedge risk converts an unnecessarily large volume of water from ownership status to rental status, thus increasing the risk associated with such water, with a resultant downgrading of the uses to which it is put. The disadvantage of a priority rights system is the higher level of transactions costs associated with matching buyers and sellers of diverse rights.

The proportional rights system appears to facilitate the organization of a market, but the type of rights cannot be separated from the way in which the return-flow issue is handled. It is necessary to recognize users' property rights in return flows if the market process is to be efficient. However, once liability for return-flow externalities is established, and since the externalities are a function of location, the uniformity of proportional rights partially disappears. Under either system, the best that can be done is to streamline the process of identifying and quantifying the return-flow effects.

The importance of return-flow externalities changes over time, especially in regions where urban growth is high. As more of the available water is used in towns where wastewater disposal is effected through a central sewer system, town-industry exchanges of rights will have no effect on the location of return flows and probably little effect on their volume.

The history of C-BT carries some interesting lessons regarding the resolution of conflicts arising from water development. Clearly, compensating the basin of origin is a necessary part of further water development in the western United States. The C-BT negotiations took several years and great expense (the Green Mountain Reservoir, for example) for success, even in an earlier era of slow economic growth and more plentiful water supplies. The competition is much keener today, and attendant environmental issues receive more attention and debate. This emphasizes that more efficient use of existing supplies will avoid or delay not only the high construction and land-opportunity costs of new projects but also the high costs of conflict resolution.

References

Anderson, R.L. 1961. "The Irrigation Rental Water Market: A Case Study," *Agricultural Economics Research*, vol. XIII, no. 2.

Anderson, Raymond L. 1965. "Emerging Nonirrigation Demands for Water," *Agricultural Economics Research*, vol. XVII, no. 4.

Asmus, E. B. 1966. "Rural-Municipal Water Transfers" (Master's thesis, Colorado State University).

Bancroft, George J. 1944. "Diversion of Water from the Western Slope," *The Colorado Magazine*, State Historical Society of Colorado, The State Museum, vol. XXI, no. 5, September 1944.

Cole, Donald Barnard. 1948. "Transmountain Water Diversion in Colorado," *The Colorado Magazine*, State Historical Society of Colorado, The State Museum, vol. XXV, no. 2.

Dille, J. M. 1958. *A Brief History of Northern Colorado Water Conservancy District and the Colorado-Big Thompson Project* (Loveland, Colo., Northern Colorado Water Conservancy District).

Dracup, J. A. 1977. "Impact on the Colorado River Basin and Southwest Water Supply," in *Climate, Climatic Change, and Water Supply* (Washington, D.C., National Academy of Sciences).

Dudley, Norman, and O. R. Burt. 1973. "Efficient Allocation of Irrigation Water Under Risk by Dynamic Programming," *Water Resources Research* vol. 9, no. 4.

Hartman, L. M., and Don Seastone. 1970. *Water Transfers: Economic Efficiency and Alternative Institutions* (Baltimore Md., Johns Hopkins University Press).

Howe, Charles W., Paul K. Alexander, and Raphael J. Moses. 1982. "The Performance of Appropriative Water Rights Systems in the Western United States during Drought," *Natural Resources Journal* vol. 22, pp. 379–389.

———, and Allan H. Murphy. 1981. "The Utilization and Impacts of Climate Information on the Development and Operations of the Colorado River System," in *Managing Climatic Resources and Risks*, Report of the Panel on Effective Use of Climate Information in Decision-Making. Climate Board, National Research Council (Washington, D.C., National Academy Press).

Kaleta, Ghebreyohannes. 1976. "Economics of Cost-Sharing Arrangements for Public Irrigation Projects" (M.S. thesis, Department of Economics, Colorado State University, Fort Collins).

Kelso, Maurice M., William E. Martin, and Lawrence E. Mack. 1973. *Water Supplies and Economic Growth in an Arid Environment: An Arizona Case Study* (Tucson: University of Arizona Press).

Northern Colorado Water Conservancy District, Annual Report (1937-1982). Loveland, Colorado.

Tipton, R. J. 1933. "Preliminary Engineering Report: Northern Transmountain Diversion" (State of Colorado, Department of the State Engineer, December).

U.S. Congress. 1937b. Senate Document 80, "Synopsis of Report on C-BT, Plan of Development and Cost Estimate," Bureau of Reclamation, Department of the Interior (Washington, D.C., U.S. Government Printing Office).

U.S. Congress, House of Representatives. 1937a. "Hearings Before the Committee on Irrigation and Reclamation on S.2681, A Bill to Authorize the Construction of the C-BT Transmountain Water Diversion Project as a Federal Reclamation Project," June 20 and July 2, 1937 (Washington, D.C., U.S. Government Printing Office).

U.S. Department of the Interior, Bureau of Reclamation. 1968. (revised). "The Story of the Colorado-Big Thompson Project" (Washington, D.C., U.S. Government Printing Office).

Index

Albright, Horace M., 179
American Forestry Association, 179
American Planning and Civic Association, 179
Anderson, R. L., 190
Anderson, W., 153
Angelides, Sotirios, 63
Arizona v. *California*, 128
Army Corps of Engineers, U.S., 41, 55, 137, 139, 141–142, 146, 151, 168
Ashworth, William, 1
Asmus, E. B., 190
Associated Engineering Consultants, 72, 83, 85, 90, 93

Bain, Joe S., 74, 76, 77
Bardach, Eugene, 63
Bauman, D., 137
Blumm, Michael C., 40n
Bonneville Dam, 26
Bonneville Power Administration (BPA), 30, 41–42
Boulder Canyon Project Act, 56, 110, 111
Bowden, Gerald D., 75
Butcher, Walter R., 33, 35
Buteau, Joanne R., 33, 35

California
  Water rights system, 75–77, 96–97
  See also Kern County, Calif.; Southern California water problem
California Aqueduct, 15, 80, 81, 82, 103, 112–113
California Department of Water Resources, 120
California State Water Resources Control Board (SWRCB), 76, 96, 97, 130
California v. *U.S.*, 54
Carey Act of 1894, 42
C-BT. *See* Colorado-Big Thompson Project
Central Arizona Project (CAP), 111–112
Central Valley Project (CVP), 16, 70, 112–113
  Surplus water from, 115, 126–127
Chaney, Ed, 42, 46
Cheung, S., 59
Clean Water Act, 145
Coase, Ronald, 58, 59
Cole, Donald Bernard, 175
Colorado-Big Thompson (C-BT) Project, 18, 171
  Construction schedule, 182–183
  History of, 175–183

Colorado-Big Thompson (C-BT)
  Project   (*continued*)
  Opposition to, 178–180
  Physical setting, 172, 173, 174, 175
  Repayment contract of 1938, 180–182
  See also Northern Colorado Water Conservancy District
Colorado Conservancy District Act, 177
Colorado River Aqueduct, 15, 103, 109–112
Colorado River Basin Project Act, 111
Colorado River Compact of 1922, 105, 109, 112, 176
Colorado River Protective Association, 178
Columbia Basin Irrigation Project, 13–14
  Expansion of, 52–53
  Political issues affecting, 56–57
  Water rights of hydropower operators, 55–56
  Water rights of irrigators, 53–55
Columbia River
  Head and generating capability on, 32, 33, 34
  Uses of, 28, 29, 31–32
  Water supply and stream flow, 26, 27, 28
Congress, U.S., 55–56
Conjunctive water use. *See under* Virginia water problem
Conley, Brian C., 116
Connecticut v. *Massachusetts*, 146
Couch v. *Clinchfield Coal Corp.*, 144
Cox, W., 143, 167
Cragwell, J. S., Jr., 149*n*
Cross Valley Canal, 80, 81, 82
CVP. *See* Central Valley Project

Davis, R. V., 154
Demand management, 3
  in Droughts, 118
  Pricing, use of, 16, 115–118, 126, 127–128
Demsetz, Harold, 59
Desalination, 3
Desert Land Act of 1877, 42
Deukmejian, George, 130
Dille, J. M., 171, 176, 178
Dracup, J. A., 176
Drought management, 118, 154–155, 159–160

Eaton, Fred, 107, 108
Economic Value of Water, The (Gibbons), vi
Ehrlich, N., 154
Elmore, John J., 120, 130
Endangered Species Act, 56
Enke, S., 92
Entitlement (firm) water, definition of, 83
Environmental Defense Fund (EDF), 123–124
Externality concerns, 10, 61–62, 97, 196–197

Federal Energy Regulatory Commission (FERC), 41, 52, 54, 55
First Iowa Hydro Electric Coop v. *FPC*, 41, 54
Fish and Wildlife Coordination Act of 1934, 42
Fisheries, 30, 56
Flood Control Act of 1944, 41
Franzini, J., 139
Frederick, Kenneth D., 7
Friant Dam, 70
Friant-Kern Canal, 70, 82

Gardner, B. Delworth, 86, 95
Geherity and Miller, Inc., 147, 148
Gershon, Sam I., 116
Gibbons, Diana C., vi
Grand Coulee Dam, 52
Green Mountain Reservoir, 178
Grimsley, J. W., 151*n*
Groundwater
  Extraction of, 70, 72, 106
  Pricing of, 86–87
  Storage of, 82–83
  Urban needs and, 146–149, 158–159
Groundwater rights
  Common law approach, 143–144
  Correlative rights, 76–77
  Reasonable-use doctrine, 144
  Reforms re, 164

Hamilton, Joel R., 44*n*, 51
Hanson, James C., 7
Hartman, L. M., 171, 197

Henningson, Durham, and Richardson, 149
Hildebrand, Carver W., 116
Hirshleifer, Jack, 74
Hoffman, Abraham, 107, 108
Holmes, B. H., 40n
Houston, J. E., Jr., 37
Howe, Charles W., 117, 176, 184
Howitt, Richard E., 73, 74, 77, 79, 92, 98
Hrezo, M., 147
Hudson, Ronald E., 151n
Hutchins, Wells A., 40n, 75
Hydropower
  Costs of, 29–30
  Fisheries, conflict with, 56
  Opportunity costs of water depletion, 34, 35
  Production, determination of, 31–32, 33, 34
  Southern California water problem and, 122–124
  Water rights of operators, 40, 49, 55–56, 59–61
  See also Irrigation-hydropower competition

Iceberg towing, 3
Ickes, Harold, 179, 180
Idaho Power Company, 12, 44, 45, 46–47
Idaho Power Company v. *State*, 48
Idaho Public Utility Commission (PUC), 46, 47
Idaho Supreme Court, 12, 48, 49
Imperial Irrigation District (IID), 16
  Competing uses for conserved water, 129–130
  Legal constraints on transfers, 128–129
  Misuse of water, 120, 121, 130
  Transfer negotiations, 131–132
  Water conservation in, 118–120, 121, 122–124
Indian water rights, 11
Interstate water allocation, 42–44
Irrigation
  New *v.* existing systems, 37–38
  Value of water and, 37–38
  Water rights re, 53–55
  See also Kern County, Calif.
Irrigation-hydropower competition, 31
  Federal law and, 40–41
  Hydropower loss, determination of, 35, 36–37
  Institutional reforms re, 58; federal and state laws, integration of, 62–63; hydropower rights subordination, elimination of, 49, 59–61; interstate externalities, 61–62
  Irrigation energy requirements, 36
  Legislative response to, 50–51
  Seasonal patterns, 35–36
  Value of water and, 37–38
  See also Columbia Basin Irrigation Project; Swan Falls case
Isabella Dam, 70
Izaak Walton League, 179

Kahrl, William L., 107, 112, 113
Kaleta, Ghebreyohannes, 194
Kelso, Maurice M., 195
Kennedy, David N., 97
Kern County, Calif., 14–15, 67, 68
  Arvin-Edison Water Storage District, 84–85
  Buena Vista Water Storage District, 85
  Contractual exchange arrangements, 83–85
  Crop values, 71, 72
  Exchange facilities in, 80, 81, 82–85
  Groundwater extraction, 70, 72
  Groundwater law, 77
  Groundwater storage, 82–83
  Imported water: allocation of, 79–80; costs of, 73; quality of, 71; sources of, 71, 72; supplies of, 70–71, 73
  Interregional transfers, 90–93
  Intraregional transfers, 93–95
  Irrigation (1800s), 69–70
  Irrigation districts, 77, 78, 79–80
  Kern Delta Water District, 85
  Land use levels, preservation of, 73–74
  Local water sources, 68–70
  Pricing and allocation rules, 85–89
  Water deficits, 90, 91
  Water needs, 67
  Water transfers, benefits of, 74, 90–95
Kern County Water Agency, 14, 70, 71, 72, 73, 77, 79–80, 82, 83–84, 90, 94
Kern River-California Aqueduct Intertie, 80, 81, 82

Lake Gaston pipeline. *See under* Virginia water problem
Lee, Clifford T., 77, 96
Lee, Kai N., 30
Letey, J., 97
Linaweaver, F. P., 117
Linsley, R., 139
Lippincott, J. B., 107
Los Angeles. *See* Southern California water problem
Los Angeles Aqueduct, 15, 103, 107–109
Lyman, Ashley, 44*n*, 51

Maguire, C. E., 155
Marysville Reservoir, 90
Masiello, R. A., 149*n*
Metropolitan Water District of Southern California (MWD), 14, 15–16, 83–84, 102, 105, 106, 109, 110, 112–118, 124–125, 128–132
Mono Lake water diversion, 108–109
Mulholland, William, 107, 108
Murphy, Allan H., 176
MWD. *See* Metropolitan Water District of Southern California

Nadeau, Remi A., 107
Nathanson, Milton N., 128
National Environmental Policy Act of 1969, 42, 145, 147
National Parks Association, 179
National Park Service, 179
NCWCD. *See* Northern Colorado Water Conservancy District
Newell, Frederick H., 107
New Jersey v. *New York*, 146
Newsweek, 1
Nor Any Drop to Drink (Ashworth), 1
North Carolina-Virginia Bi-State Water Resources Management Committee, 145, 148–149
Northern Colorado Water Conservancy District (NCWCD), 171
 Boundaries of, 172, 173
 Establishment of, 177–178
 Water markets: allocation rules, 183–185, 197, 198; allotment markets, 187–189; alternate supply costs and, 192, 193, 194–195;
 effectiveness of, 22–24; flexibility of, 195; price data, 188–189, 190–191; proportional rules re, 23, 184, 197, 198; rental markets, 189–192; return flow provisions, 18, 23, 181–182, 185–186, 196–197, 198; speculation in, 194–195; "tradable margin" of water supply, 196
 See also Colorado-Big Thompson Project
Northern Colorado Water Users Association, 177
Northwest Power Planning and Conservation Act, 56
Norwood, Gus, 40*n*, 41
Nuclear power projects, 30

Opportunity costs, 3, 34, 35
Owens Valley water diversion, 107–108

Pacific Northwest
 Hydropower in, 29–30
 Interstate water allocation, 42–44
 Irrigation in, 28–29
 Opportunity costs of water depletion, 34, 35
 Water agencies in, 41–42
 Water rights in, 39–40
 See also Columbia Basin Irrigation Project; Swan Falls case
Pacific Northwest Power Planning Council (NPPC), 42
Petke, Daniel L., 26
Phelps, Charles E., 79, 86, 95, 96, 97
Powledge, Fred, 1
Prinderville, S., 140*n*, 141*n*
Prudent management yield of a water system, 140–141
Pyle, Stuart T., 87

Raiffa, H., 166
Reclamation, U.S. Bureau of, 40–42, 52, 53, 62, 70, 94, 112, 120, 175, 176, 177
Return flows, ownership of, 18, 23, 181–182, 185–186, 196–197, 198
Richman, R., 166

INDEX

205

Rivers and Harbors Act of 1899, 41
Rogers, Peter, 6
Roosevelt, Franklin D., 180
Roosevelt, Theodore, 108

Schelhorse, Larry D., 116
Schuy, David F., 35
Science 81, 1
Seastone, Don, 171, 197
Second National Water Assessment, v
Seven-Party Water Agreement, 110
Shabman, L., 134, 143, 165, 167
Sheer, D., 134, 153n, 154
Snake River
  Head and generating capability on, 32, 33, 34
  Uses of, 31–32
  Water supply and stream flow, 26, 27, 28
  See also Swan Falls case
Solley, Wayne B., 6
Southern California water problem, 5, 15–16
  Groundwater pumping, 106
  Hydropower generation and, 122–124
  Irrigated areas, 102, 103
  Local water supplies, 105–106
  Mono Lake, water from, 108–109
  Owens Valley, water from, 107–108
  Supply and demand alternatives: demand management, 115–118, 126, 127–128; Imperial Irrigation District, water from, 118–120, 121, 122–124, 128–132; State Water Project additions, 114–115; surplus water purchases, 115, 126–127
  Water deficit, projected, 113
  Water import systems, 102–103, 104, 105
State of Washington Department of Game v. FPC, 41n
State Water Project (SWP), 14, 71, 83, 104, 105, 113, 114–115
State Water Resources Development Bond Act of 1959 (Calif.), 113
Stoutemyer, B. E., 49
Supreme Court, U.S., 11, 41, 43, 45, 54–55, 61, 146
Supreme Court in California v. the United States, 41n

Surplus water, definition of, 83
Swan Falls case, 12–13
  Background, 44–46
  Court decision re, 48
  Legislative response, 49–51
  Water rights conflict, 46–51
SWP. *See* State Water Project

Taylor, Edward T., 177
Thermal power, 29–30
Tipton, R. J., 175, 176–177
Town of Gordonsville v. Zinn, 144
Town of Purcellville v. Potts, 144, 145
Transmountain diversions. *See* Colorado-Big Thompson Project
Trelease, Frank J., 58

U.S. News and World Report, 1
U.S. v. New Mexico, 54
U.S. v. Rio Grande Dam & Irrigation Co., 55
Urban water needs. *See* Southern California water problem; Virginia water problem
Utton, A. E., 43

Vaux, H. J., Jr., 74, 92, 95, 97
Virginia Groundwater Act of 1973, 143–144, 147, 162, 164
Virginia Hot Springs v. Hoover, 144
The Virginian-Pilot, 17
Virginia State Water Control Board, 139, 141, 147, 148, 149, 151, 162, 164, 168
Virginia Supreme Court, 144
Virginia water problem, 5, 20, 134–135
  Conjunctive water use alternative, 17; construction costs, 155–156; efficiency of, 153–155; excess capacity concerns, 161; legal challenge concerns, 160; operating costs, 155–156
  Federal government, role of, 145–146
  Groundwater alternative, 146–149
  Intercommunity distrust, 150
  Interstate institutional mechanisms, role of, 145, 146, 148–149
  Lake Gaston pipeline, 17–18, 142; construction costs, 155–156;

Lake Gaston pipeline (*continued*)
consumer water bills and, 157–158; decision to implement, 149–150; drought management and, 159–160; efficiency of, 152–153; excess capacity concerns, 161; groundwater conflicts and, 158–159; legal challenge concerns, 160; operating costs, 155–156; opposition to, 151–152; reservoir, effect on, 151; selection criteria, 158–161
Local government, role of, 143
Problem region, 135, 136
State government, role of, 143–144
Water deficit, 135, 137, 141–142
Water transfers alternative, 161–162; groundwater rights, reforms re, 164; negotiation costs, 163; Permit Authority, proposed, 165–168; technical information re, 163–164; water markets, 164–165; water rights concerns, 162–164
Water use projections, 137–138, 139
Yield estimates, 139–141

Walker, David, 33, 35
Washington Public Power Supply System, 30
Water: The Nature, Uses, and Future of Our Most Precious and Abused Resource (Powledge), 1
Water banks, 51, 63
Water crisis, 1–2
Water management, efficiency in, 4–5
Water markets, 171–172
Allocation rules: priority system, 183–184; proportional system, 23, 184, 197, 198; quota system, 185
Allotment markets, 187–189
Alternate supply costs and, 192, 193, 194–195
Effectiveness, reasons for, 22–24
Externalities and, 10, 196–197
Federal role in, 22
Flexibility of, 195
Institutional deficiencies re, 9–10
Obstacles to, 21–22
Price data re, 188–189, 190–191
Rental markets, 189–192
Return flows, provisions re, 18, 23, 181–182, 185–186, 196–197, 198
Speculation in, 194–195
"Tradable margin" of water supply, 196
for Urban needs, 164–165
Water rights and, 9
Water pricing, 85–86
Average cost pricing, 86
Average prices, 87, 88
Cost pricing, 7
in Demand management, 16, 115–118, 126, 127–128
Groundwater *v.* surface water, 86–87
Rationing function, 87–89
Variable charges, 86
in Water markets, 188–189, 190–191
Water project construction, federal regulation of, 145–146
Water resource development, 2–3
Water rights
Appropriative rights, 8, 39–40, 75–76, 77, 96
Common law approach, 143–145
Correlative rights, 76–77
Equitable apportionment, 43, 146
Federal and state laws, integration of, 43–44, 52–57
Federal laws, 40–43, 55–56
Federal reserved rights, 11
Flow and diversionary uses, equal applicability to, 48
Groundwater rights, 76–77, 143–144, 164
of Hydropower operators, 40, 49, 55–56, 59–61
Indian rights, 11
Interstate allocation and, 42–44, 52–57
of Irrigators, 53–55
Outdated nature of, 8–9
Prescriptive rights, 75
Reforms re, 20–21, 43–44, 52–57
Riparian rights, 8, 75, 77, 96, 144–145
State laws, 7–8, 39–40, 43–44, 52–57
in Swan Falls case, 46–51
Water markets and, 9
Water transfers and, 96–97, 162–163
Water shortages, fears re, 19–20
Water transfers, 15, 74
Benefits of, 89–95
Contractual exchange arrangements, 83–85

Elevation gradient and, 93
Exchange facilities, 80, 81, 82–85
Externalities, concerns re, 61–62, 97
Flexibility in, 21
Import demand, determination of, 91–92
Incentives re, 20
Institutional reforms re, 58–63; federal and state laws, integration of, 62–63; hydropower rights subordination, elimination of, 49, 59–61; interstate externalities, correction of, 61–62
Interregional trade, 90–93, 97–99
Intraregional trade, 93–95
Legal barriers, 95–99, 128–129
Permit Authority, proposed, 165–168
Price gradient and, 94–95
Prices, establishment of, 89
Profits from, 98
for Urban needs, 118–120, 121, 122–124, 161–162; competing uses for water, 129–130; federal and state support, 130; groundwater rights, reforms re, 164; legal constraints, 128–129; negotiating the transfer, 131–132, 163; technical information re, 163–164; water markets, use of, 164–165
Water rights and, 96–97, 162–164
See also Water markets
Water use patterns in the U.S., 6–7
Weather modification, 3
Whitman, Requardt, and Associates, 149, 154, 156
Whittlesey, Norman, 33, 35, 36, 37, 38, 53
Wild and Scenic Rivers Act of 1968, 42

Yield of a water system, 139–141
Yuba County Water Authority, 90

CARROLL COLLEGE LIBRARY
HELENA, MONTANA 59625